Exploring Math with Technology

This timely book provides support for secondary mathematics teachers learning how to enact high-quality, equitable math instruction with dynamic, mathematics-specific technologies.

Using practical advice from their own work as well as from interviews with 23 exceptional technology-using math teachers, the authors develop a vision of teaching with technology that positions all students as powerful doers of mathematics using math-specific technologies (e.g., dynamic graphing and geometry applications, data exploration tools, computer algebra systems, virtual manipulatives). Each chapter includes sample tasks, advice from technology-using math teachers, and guiding questions to help teachers with implementation. The book offers a rich space for secondary math teachers to explore important pedagogical practices related to teaching with technology, combined with broader discussions of changing the narratives about students – emphasizing the mathematics they can do and the mathematics they deserve.

Accompanying online support materials include video vignettes of teachers and students interacting around technology-enhanced tasks in the classroom, as well as examples of more than 30 high-quality technology-enhanced tasks.

Allison W. McCulloch is Professor of Mathematics Education in the Department of Mathematics and Statistics at the University of North Carolina Charlotte, USA.

Jennifer N. Lovett is Associate Professor of Mathematics Education in the Department of Mathematical Sciences at Middle Tennessee State University, USA.

W0113056

This extraordinary volume features meaningful, practical tools to help you integrate technology into your classroom in ways that position your students as mathematical explorers. Rich, high-cognitive demand tasks and vignettes using those tasks in classrooms help you dig deep into teaching mathematics with technology in ways that support meaningful student engagement. The authors make connections to practices and frameworks like the 5 Practices that will help you seamlessly integrate new and innovative technology into your current mathematics teaching practices.

– **Mike Steele**, Past President, *Association of Mathematics Teacher Educators*

McCulloch, Lovett, and the 23 brilliant secondary teachers showcased in the book are thought leaders in guiding educators on how to integrate technology in the math classroom. Positioning students as math explorers captures the essence and vision of what equitable teaching should look like when using technology to "remove barriers and provide students agency". As they share engaging exemplar lessons, vignettes with student work, and videos, educators can learn how to bring this vision to life in their own classrooms so that each and every student can use dynamic technology to build insights, deeper conceptual understanding, and confidence as mathematicians. A must-have book for all educators!

– **Jennifer Suh**, Professor of Mathematics Education, *George Mason University*

This is a really rare examination of teachers, students, math, and technology, one with a keen understanding of their value individually but also the ways that thoughtful pedagogy can multiply that value.

– **Dan Meyer**, Director of Research, *Desmos Classroom*

Exploring Math with Technology

Practices for Secondary Math Teachers

Allison W. McCulloch and Jennifer N. Lovett

Routledge
Taylor & Francis Group

NEW YORK AND LONDON

Designed cover image: © Getty Images

First published 2024
by Routledge
605 Third Avenue, New York, NY 10158

and by Routledge
4 Park Square, Milton Park, Abingdon, Oxon, OX14 4RN

Routledge is an imprint of the Taylor & Francis Group, an informa business

ISBN: 978-1-032-29838-2 (hbk)
ISBN: 978-1-032-29837-5 (pbk)
ISBN: 978-1-003-30228-5 (ebk)

DOI: 10.4324/9781003302285

Typeset in Palatino
by SPi Technologies India Pvt Ltd (Straive)

Access the companion website: https://www.tlmtresearch.com/teachingmathtechbook

Dedication

This book is dedicated to Ian, Brendan, Gavin, and Jenn who always lift me up and remind me that I have something to share.

~ Allison

This book is dedicated to Kyle and Claire who were unendingly patient when I was writing rather than playing.

~ Jen

Contents

Links

You can find links to all the technology-enhanced tasks and supplementary videos throughout the book at https://www.tlmtresearch.com/teachingmathtechbook.

Acknowledgments

This book was born out of a long career of teaching math with technology and working with others learning to do so. We have had the privilege of thinking about both student and teacher learning in the context of math-specific technologies with amazing teachers and colleagues over the years. For Allison, that started with Max Coleman and Mark Klespis, who introduced her to the TI-81 and 92 and changed her pedagogy forever. For Jen, that started with Suzanne Harper, who introduced her to Geometer's Sketchpad. Together, we have worked with and learned from many math teacher educators over the last 15 years. (Special shout-out to Hollylynne Lee and Karen Hollebrands!) We would also like to thank Lara Dick and Nina Bailey for their unending support, read-throughs, and advice that were invaluable throughout this process.

This book would not have been possible without the input of the 23 outstanding technology-using secondary math teachers who shared their expertise with us. We spent time with each of them learning about such things as how they decide when to use technology in their instruction, the ways they set up norms and routines in their classrooms around technology use, and how they identify and use technology-enhanced tasks in their instruction. They were all so generous with their time and excited to share what they have learned about incorporating technology-enhanced math tasks into their teaching to help others do the same. We are so grateful for their time and thoughtfulness. A special thank you to Kristen Fye for inviting us into her classroom over the course of a year to observe and capture her practice when using technology-enhanced math tasks firsthand.

You will hear directly from these teachers throughout this book. Since calling them the *Technology Using Secondary Math Teachers* is simply too much to keep repeating, they will heretofore be referred to as the *Tech-Math Teachers*. They are all introduced next.

Meet the Tech-Math Teachers

Dan Anderson*
Dan is a high school math and computer science teacher in Clifton Park, New York. He is also a Desmos Fellow who enjoys finding the commonalities between the math and computer science curricula. Dan's favorite part about using technology in math class is using it to enable students to express their mathematical creativity and explore ideas that would be impossible without tech.

Nina Bailey*
Nina is a former high school math teacher who now teaches introductory statistics and courses for future high school math teachers and works with high school students in after-school math sessions. She loves how math action technologies provide opportunities for her students to explore and engage in sensemaking about important mathematical and statistics concepts before she introduces them to more formal language and procedures.

Lauren Baucom*
Lauren is currently a curriculum developer for Amplify Desmos Math. A former high school mathematics teacher of 12 years, Lauren has devoted her career to helping students from both rural and urban areas experience mathematics in a way that teaches them about themselves, others, and their world. As a recent doctoral graduate, in her research, she prioritizes how to use the teaching of statistics with technology as a tool to disrupt power structures in the math educational system by seeking justice for students and teachers of marginalized populations.

Joel Bezaire*
Joel teaches eighth-grade Algebra 1 at the University School of Nashville in Nashville, Tennessee. He is a Desmos Fellow and Certified Presenter and a Rosenthal Prize finalist, and he serves on the board of the Educators' Cooperative. Joel enjoys using Desmos and CODAP to help students achieve a deep, conceptual understanding of tricky topics, particularly in algebra and data analysis.

Nick Corley

Nick is a middle school math teacher in Northfield, New Jersey, with more than 20 years of experience in the classroom. In addition to teaching, he is a Desmos Fellow, a Desmos Certified Presenter, an Amplify Professional Learning Specialist, and a curriculum consultant. Nick enjoys talking with and supporting other math teachers by using edtech tools to energize math instruction.

Nolan Doyle

Nolan is a high school math teacher in Chesterfield, Virginia. He is also a Desmos Certified Presenter who loves sharing his experiences from both the fellowship and his classroom practice with other educators at local, state, and national conferences. He continues to find ways to support teachers in using the Desmos tools to help every student learn math and love to learn math.

Samantha Fitzgerald

Samantha has been teaching high school math in North Carolina for four years. She first learned about math-specific technology tools when completing her undergraduate degree at the University of North Carolina at Charlotte. Since then, she has incorporated tools like Desmos and GeoGebra into her teaching. Her favorite thing about teaching with technology is the collective participation in discussions it can create as well as providing access for nonnative English speakers.

Karen Forrester

Karen teaches high school math in Des Moines, Iowa. Both Desmos and GeoGebra help her students reach a deep understanding of math concepts. Technology helps them see patterns, make connections, and support their reasoning.

Sarah Furman

Sarah taught English for 10 years and then mathematics for 13 years in rural northern Michigan. She is a Desmos Fellow, a member of the Open Middle team, and a participant in the Rosenthal Prize Institute for math educators at the National Mathematics Museum. She is passionate about making the mathematics classroom a more equitable, playful, and creative environment for students.

Kristen Fye*

Kristen teaches high school math and an early college in Charlotte, North Carolina. She loves to use technology tools that provide her students an opportunity to be curious and build on their intuitions and prior experiences. She is a Desmos Fellow and is currently a mathematics education doctoral student. She is especially interested in the ways that engaging in technology-enhanced math tasks can support the development of students' positive and powerful math identities.

Juan Gomez*
Juan is a high school teacher in Carmel, California. He is a Desmos Fellow and AP Statistics reader who enjoys using technology to spark curiosity in his high school students.

Andy Hamric*
Andrew is a high school teacher that has been teaching math in the Tuscaloosa, Alabama, area for 17 years. He was introduced to teaching with technology in his undergraduate education program and enjoys using it in his class. He believes that using technology to teach mathematics gives students access to math ideas that they might otherwise find too difficult or time-consuming.

Shauna Hedgepeth*
Shauna has been teaching secondary math and science in Mississippi for more than 20 years and has incorporated technology throughout her teaching career. She loves working with students as they become curious about and engaged in mathematics through the use of technology. Her passion, however, is helping students have a healthy skepticism of the statistics they consume in their use of social media.

Kristy Jacob
Kristy is a high school mathematics teacher at Riverview East Academy, located in Cincinnati, Ohio. She is a fifth-year teacher who is passionate about exploring the intersection between math and art to help make mathematics visual and accessible for all learners.

Nathan Kenny
Nathan has been a high school math teacher in Alabama since 2013. He first learned about mathematical action technologies (MATs) after teaching for a few years and while pursuing an alternative master's degree in secondary math education. He found that MATs like Desmos, GeoGebra, and TI-Nspires empower students in building strong conceptual understanding through exploration and discovery. As a National Board–certified teacher and a fellow in the National Science Foundation funded A-PLUS in Math Program at the University of Alabama, he continues to grow in effectively implementing

such technologies to improve mathematical discourse and to improve students' positive mathematical identities, self-efficacy, and agency.

Allyson Klovekorn

Allyson is a sixth-year high school math teacher in Oxford, Ohio. She enjoys making connections with her students and helping them grow and reach their goals. She also loves incorporating new activities that utilize technology, especially Desmos and GeoGebra, to help enhance her students' understanding of vocabulary and concepts.

Sabrina Monserate

Sabrina is a high school math teacher in Durham, North Carolina. She first learned about math-specific technology tools when completing her undergraduate degree at NC State University. Since then, she has incorporated tools like Desmos, GeoGebra, and CODAP into her teaching.

Zachary Patterson

Zack is a math teacher and instructional coach at New Albany High School near Columbus, Ohio. Zack is passionate about making math meaningful for students. He utilizes technology to help foster and support students' natural curiosity to engage with mathematics in the classroom.

Andrew Schwartz*

Andrew is a high school math teacher in northeastern Colorado. He has used math technology (such as Desmos and GeoGebra) in Geometry, Algebra 2, Precalculus, and Calculus 1. He likes using technology that allows students to explore mathematical concepts and visualize abstract ideas.

Leah Simon

Leah is a computation layer specialist on the Desmos Classroom team at Amplify. Prior to that, she taught high school math in Dayton, Ohio. She is passionate about creating digital lessons that provide students with a variety of ways to explore, make sense of, and deepen their understanding of mathematics.

Michele Torres

Michele is a high school teacher at Community High School in West Chicago, Illinois. She is a Desmos Fellow and likes to share ways to use dynamic geometry and other technologies with other teachers. She enjoys finding creative ways to help students discover mathematical concepts using technology.

Harsh Upadhyay

Harsh is a high school mathematics teacher in Louisville, Kentucky. He is also a Desmos Fellow who works with other secondary teachers learning to teach math with technology. He has found that dynamic technologies really help his students make connections among representations in ways that are otherwise difficult to achieve.

Mark Vasicek

Mark teaches at Burlington high school in Iowa. He uses Desmos and Geogebra to support students as they explore math. With dynamic geometry technology, students are not afraid to make mistakes; when they do, they persevere in trying again.

Drew Willett

Drew teaches high school math in North Carolina. He learned about teaching with technology-enhanced tasks during his undergraduate program at the University of North Carolina at Charlotte. He regularly uses tools like Demos and CODAP and likes how they can support the development of conceptual understandings.

* A special thank-you to these teachers for sharing some specific technology-enhanced math tasks along with detailed plans that include anticipations of student work, how they launch the task, what they look for when monitoring student work, and how they plan for orchestrating a whole-class discussion.

About the Authors

Allison W. McCulloch is a professor of mathematics education at the University of North Carolina at Charlotte in North Carolina, where she teaches mathematics, mathematics education, and research courses for future secondary mathematics teachers and graduate students. Prior to working with prospective teachers, Allison spent over a decade teaching middle and high school mathematics. As a secondary mathematics teacher in the 1990s, she was an early adopter of math action technologies to support her students' learning and her insight into their thinking. As a mathematics teacher educator, she focuses her work largely on supporting and studying the teaching and learning of mathematics with technology and designing curricula for teachers learning to teach secondary mathematics with technology.

Jennifer N. Lovett is an associate professor of mathematics education at Middle Tennessee State University, where she teaches mathematics/statistics, mathematics education, and research courses for future secondary mathematics teachers and graduate students. Prior to working with prospective teachers, Jennifer was a middle school and high school mathematics teacher. As a mathematics teacher, she regularly incorporated math action tools into her teaching to support all her students. Jennifer has extensive experience supporting and studying the teaching and learning of mathematics with technology and creating curricula for teachers learning to teach secondary mathematics with technology.

Part I

Planning for and Implementing Technology-Enhanced Math Tasks

My (Allison's) personal journey with teaching with technology began as a high school teacher in the early 1990s with a class set of TI-81 graphing calculators (yes, that long ago!). All of a sudden, we could not only complete computations but also create graphs quickly and accurately. More important, however, having this tool changed the ways that I could have my students engage with many math concepts. Using the numerical computations, tables, and graphs we created with the graphing calculator, we could explore, look for patterns, and test our conjectures. Positioning the tool in this way (as something you could use at any time to explore a new idea) changed who had access to the math discussions in my classroom. It invited all students (regardless of their prior academic success) to think in mathematical ways and feel welcome in our mathematical space. Those early experiences solidified my belief that technology tools, when positioned appropriately, can profoundly impact students' math experiences.

My (Jennifer's) personal journey with teaching with technology began in my educational math technology course during my teacher preparation program. On the first day of class, I wrote in a reflection to my professor that I believed that "technology took away from the understanding of mathematical concepts and the trouble of dealing with technology does not outweigh its benefits". At this time my only experience with technology in my math classes was using a graphing calculator and not in the ways Allison described. However, after learning about the ways that dynamic geometry and statistical software can support students' meaning-making about important big ideas, I left

DOI: 10.4324/9781003302285-1

this course on the last day believing that technology was a "crucial" aspect of the classroom and I wrote in a reflection that I believed "it allowed students to explore some math concepts in ways that weren't possible with paper and pencil". This belief followed me into the classroom, where I became a relatively early user of dynamic geometry software. Incorporating this software allowed all my students a way to explore a geometric theorem and begin to develop conjectures and informal proofs. The power of those classroom experiences firmly established my beliefs in the power of technology to provide students with these informal experiences.

We've come a long way from our first experiences with using technology to teach math. This has included learning to use different technology tools (Geometer's Sketchpad, Cabri, SimCalc, TinkerPlots, Fathom, Virtual Manipulatives, GeoGebra, Desmos, and more) as they have been developed. No matter our starting points, we each had to learn on our own until we found like-minded colleagues with whom to collaborate. We both followed the path of being called on to help others in our schools and districts learn to teach with technology, and eventually ended up in the same place at the same time and got to collaborate in thinking about how to support others on a larger scale. For years this was focused on learning the technologies themselves. We have hit a point in time at which most teachers we encounter are using technology in some way, whether it is graphing calculators or free web-based tools like Desmos. There are many resources available to support teachers in learning how to use these tools. However, there are very few resources that help teachers think about how technology fits into the picture of everything else we know about good teaching. For us, this includes providing opportunities for each and every student to access and explore mathematics, communicate their mathematical ideas, and develop informal and powerful ways of thinking mathematically. In writing this book, we hope to ensure that as technology is making its way into more and more math classrooms, it is being used in ways that position all students as doers of mathematics.

What We Mean by Technology-Enhanced Math Tasks

When you think about using technology in secondary math classrooms, many different images might come to mind – projectors displaying images or slides, students using graphing calculators, students accessing information on their phones, laptops out on every desk with students engaging in activities that use dynamic technologies, completing online practice problems, or watching math videos. All of these technologies generally fall into two categories – conveyance and what some refer to as "math action" technologies (Dick & Hollebrands, 2011). Conveyance technologies are just like they sound, they convey information. They allow students and teachers to present information, communicate, and collaborate with each other (e.g., PowerPoint, Google Docs, video).

While conveyance technologies are very important in the math classroom, the focus of this book is going to be on the use of *math action technologies*. (From this point on, when we use the term *technology*, we are referring to math action technologies.) These are technologies that respond to user actions in mathematically defined ways (Dick & Hollebrands, 2011). For example, if a student drags the vertex of a triangle that is constructed using dynamic geometry technology, the vertex will not only move, but the image will also keep mathematically defined relationships intact according to how the triangle was constructed, allowing the student to see how the construction responds to their actions (Figure 0.1). Or if a student enters "$y = mx + b$" into a dynamic graphing calculator and creates sliders for m and b, the technology will respond by creating a graph of the line with the slope and y-intercept defined by the sliders. As the sliders are dragged, the values of m and b change accordingly allowing the student to see how the graph of the line responds to their actions (Figure 0.2). These types of technologies allow students to interact with mathematical objects in ways that are difficult or not possible with pencil and paper. Since technologies of this type drastically change the ways in which students can interact with mathematical objects and express their thinking about them, integrating them into instruction can

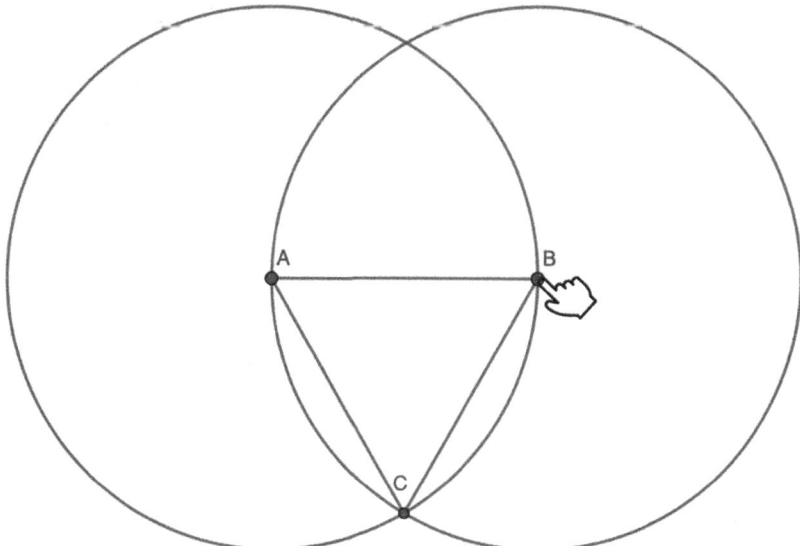

Figure 0.1 A triangle constructed in a dynamic geometry program.

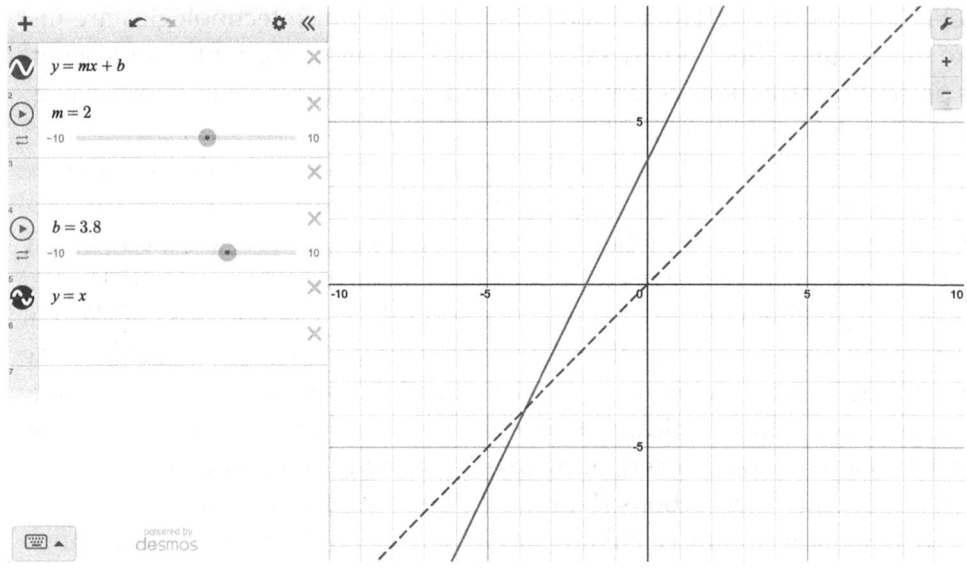

Figure 0.2 Sliders to explore a linear function in a dynamic graphing program.

be a challenge. That is where this book comes in. We draw on interviews with 23 technology-using secondary math teachers and over two decades of research to share what we have learned about teaching math with technology.

So what does it mean to teach math *with* technology effectively? The word *with* suggests that the technology plays a key role in supporting students' learning. Effective instruction can be thought of as a collection of interactions among students and teachers around the content and tasks, sometimes referred to as the instructional triangle (Cohen et al., 2003; Figure 0.3 on the left). When incorporating technology with a task, as either an integral part of the task or as a tool alongside it, we can think of effective instruction being the interaction among students, teachers, and tasks + technology (Figure 0.3

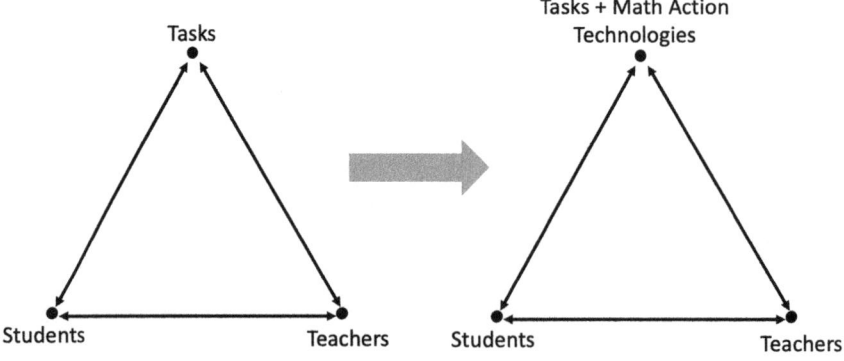

Figure 0.3 The "instructional triangle" and the "instructional triangle with technology-enhanced tasks".

on the right). Importantly, the technology itself does not teach math. That is why throughout the book you will see a focus on these interactions – how to plan for them and how to facilitate them.

What You Can Expect in This Book

Now that you know the types of technologies we will focus on in this book and how we think about effective instruction with them, you are ready to dig in and learn how to plan for and implement technology effectively in your classroom. You will hear not only from us but also from the outstanding teachers, the Tech-Math Teachers, who have shared their practice and insight (see the Acknowledgments). Here is what you can expect from this book:

- Advice (and supplemental video) from the Tech-Math Teachers on their decision-making related to planning for and reflecting on their teaching with technology-enhanced math tasks
- Vignettes (and supplemental video) of authentic student work on technology-enhanced math tasks and whole class instruction with a technology-enhanced math task in real secondary classrooms
- A set of "questions to discuss with colleagues" to encourage collaboration.
- A summary of guiding questions to support the implementation of the big ideas at the end of each chapter

 You can find links to all the technology-enhanced tasks and supplementary videos throughout the book at https://www.tlmtresearch.com/teachingmathtechbook.

1

Using Technology to Position Students as Math Explorers

When you think about students engaging with technology to support their math learning, what do you envision? What are students doing? How are they using the technology? We find it most helpful to begin thinking about effective and equitable teaching with technology by focusing on students' experiences. We love the way that Francis Su, in his wonderful book *Mathematics for Human Flourishing* (2020), talks about doing math – math is exploring. That is how we like to think about students – as math explorers. Explorers ask questions like "Why?" "How?" "What if?" and "Always?" When we think about students in this way, then we think about technologies as the tools with which they can explore!

If we think about students as explorers, that means they should be engaging with tasks + technology that foster their curiosity and provide ways for them to interact with and act on mathematical objects to test their ideas. Let's look at an example. Figures 1.1 and 1.2 are images from two different technology tasks that are focused on the same math concept, comparing measures of center. In the first example (Figure 1.1), students are asked to compare the measures of center of two static data distributions and determine if they are similar or different. They use the technology to answer the question and will see if they are right or wrong. In the second example (Figure 1.2), the students are asked to drag the points in the two different dot plots so that they have the same mean, 6. As they drag the points, the values on the left (mean and median) change accordingly. Both of these activities use technology, they are focused on the same content, and both provide feedback to students

DOI: 10.4324/9781003302285-2

Comparing center and spread

⊞ Google Classroom

The Baltimore Ravens won the American football championship in 2013. In the same season, the Kansas City Chiefs had a record of 2 wins and 14 losses.

The dot plots below show the 40-yard dash times (in seconds) of the defensive backs on each team's roster. Defensive backs mainly defend against passes, and the 40-yard dash is the main measure of speed used in American football.

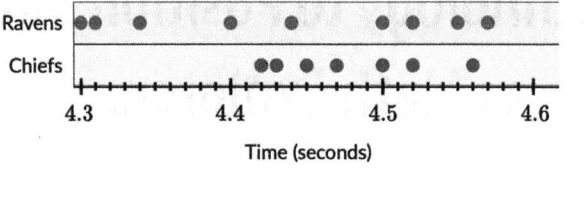

The center of the Ravens distribution is [⌄] the center of the Chiefs distribution.

The variability in the Ravens distribution [✓] the Chiefs distribution.

> greater than
>
> less than
>
> about the same as

Stuck? Use a hint.   Report a problem

Figure 1.1 Comparing measures of center example 1.

Challenge: Drag the points to create two different data sets that both have a mean of 6.

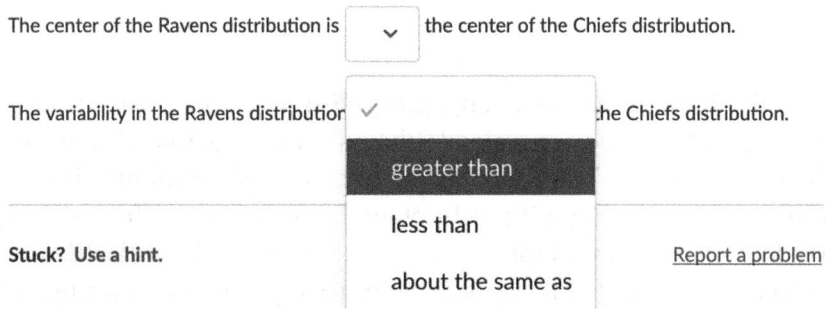

Figure 1.2 Comparing measures of center example 2.

depending on how they respond. However, only the second positions students as math explorers. It allows students to drag all the points in the dot plot and see the consequences of their actions. In that way, they can ask and explore questions like "Why does changing the point on the far right change the mean but not the median?" "How can I get the mean to be the same for both data sets?" "What if I only change the points near the middle of each plot?" and "Is it always possible to create data sets with different data but the same mean?" In asking and exploring questions such as these, students' understanding of what the mean is (a balance point) and how that relates to how we calculate it (mean $= \dfrac{\text{sum of the terms}}{\text{number of terms}}$) will inevitably deepen.

We will be emphasizing this notion of students as explorers throughout the book. We do this because we have found that thinking of students and math in this way helps us address the inequities that often exist in math classrooms, especially where technology is concerned. This includes that many students only get to use technology to practice procedures or to engage with remediation software programs and that it is often only certain students, those that work quickly or have been tracked into "advanced" classes, that get to engage with technology that positions them as explorers. We feel very strongly that it is important that all students get to use technology in ways that support their development as doers of mathematics. We hope by referring to them as explorers will help you think about that instructional triangle (see Figure 0.3) in a very specific way – as the interactions among teachers and students, as they explore math through tasks + technology.

When deciding whether a technology-enhanced task positions students as math explorers, we always look for the ways it positions all students to do the following:

1. Access and explore the math
2. Communicate their mathematical ideas
3. Develop informal and powerful ways of thinking mathematically

A really good technology-enhanced math task, like any good math task, has a way for every student to enter the problem. That is what we mean by positioning students to *access and explore the math*. In other words, ask yourself if there is a way for every student to access and engage in mathematical exploration using the technology? These kinds of tasks are sometimes referred to as low floor and high ceiling, meaning the floor is at a level where everyone can walk in and the ceiling is high enough that there are lots of possibilities to explore even further (Papert, 1993). We can also consider the walls of a task (Resnick & Silverman, 2005). A task with wide walls provides space to explore the math through multiple pathways. Sometimes it is the use of technology that opens up the walls of a task, allowing students additional pathways from the floor to the ceiling.

Consider the second Comparing Measures of Center task in Figure 1.2. Every student can begin exploring by simply dragging points in the dot plot and observing what happens as a result – low floor. Because the dot plot and display of statistical measures (mean, median) are dynamically linked, there are many different actions students can take to determine a solution (and there is more than one solution) – wide walls. Finally, because students are positioned as explorers, they can ask all sorts of additional questions and explore them – high ceiling.

When engaging in mathematics with others, students communicate their mathematical ideas in many different ways. For example, they talk about their thinking, draw pictures, write words, and create equations. When working with a technology-enhanced task, the technology affords additional ways that students can express their mathematical ideas. They can create dynamic images through dragging to help share their thinking, they can point to objects they have created that they do not yet have words to describe to express their ideas, and sometimes we can get a sense of their thinking by watching what they do with the technology without saying any words. That is what we mean when we say the technology positions students to *communicate their mathematical ideas*. Looking back at our earlier example, you can imagine students discussing their strategies for creating a mean of 6 for both data sets, and if we listen to only what they say, we might hear one student say, "What if we try this? No, that didn't work. How about if we drag it here?" and the other then add, "Look, we're getting closer! I think it will change the mean more if we drag it that way". These students are clearly sharing important mathematical ideas with each other. These ideas are not only being partially expressed through their words, but also their actions with the technology are a critical part of their mathematical communication. Imagine how important this would be if the students do not both speak the dominant language in the classroom! Just like images can support communication in those situations, the dynamic images students create when interacting with technology can support the communication of mathematical ideas when students' ability to use the dominant language in the classroom, or even formal mathematical language, is developing.

As students have opportunities to explore mathematics and use what they are creating with technology to help communicate their mathematical ideas they are being positioned to develop *informal and powerful ways of thinking* mathematically. Once again, consider the Comparing Measures of Center example, as the students drag a point in the data set to the far right, they are seeing the mean change in a very specific way as they do so – it is pulling away from the median and getting larger as they drag it farther. They are developing really important ways of thinking about the mean and median. For example, they may be developing ideas about how outliers affect the different measures of center and ideas about which measures of center are resistant to outliers. They may not know what outliers are or what a skewed data set is yet, but they are developing very important ideas about them. The teacher can then support students by introducing mathematical terms to help them express their powerful ideas in mathematically precise ways.

The ideas we are presenting here are not new. For years we have known that equitable instructional practices rely on these components. There is a need for complex tasks that support meaningful problem-solving, the use of tools to represent mathematical ideas, and classroom norms focused on students sharing their own understanding in classrooms to promote equitable instruction (including those classrooms with English language learners and students with disabilities; Carpenter & Lehrer, 1999). When these instructional ideas first took off, technology in the math classroom was nothing like what is available today. Now that such technologies are readily available to students, it is important to carefully consider how we position them in instruction. This is articulated in the National Council of Teachers of Mathematics' (NCTM) essential elements of mathematics instruction, which states,

> An excellent mathematics program integrates the use of mathematical tools and technology as essential resources to help students learn and make sense of mathematical ideas, reason mathematically, and communicate their mathematical thinking.
>
> (NCTM, 2014, p. 78)

Let's look at some examples of students engaged with technology-enhanced math tasks that positioned them as math explorers – providing them ways to access and explore the math, communicate their mathematical ideas, and develop informal and powerful ways of thinking mathematically.

Example 1: Tangent Line to a Circle

When introducing students to new mathematical vocabulary and relationships it is helpful to have them explore those relationships by acting on the

Explore (drag the points): What do you notice? What do you wonder?

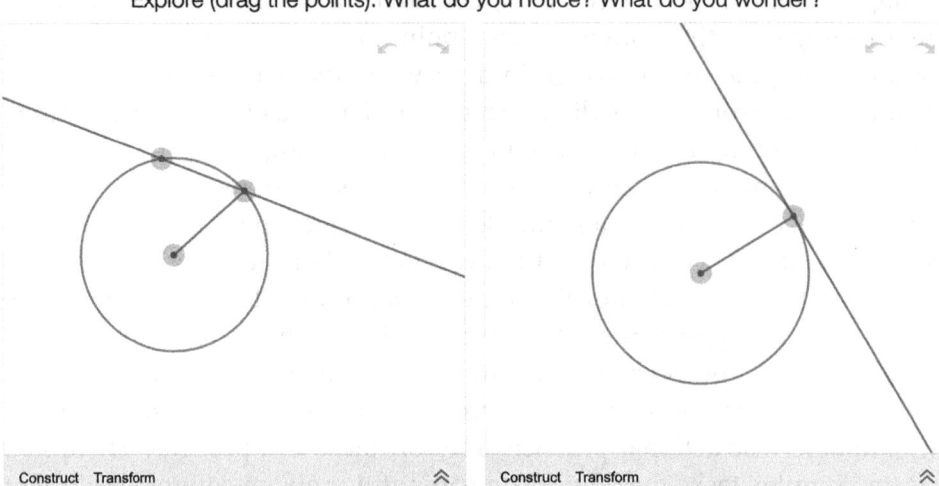

Figure 1.3 Exploring secant and tangent lines using dynamic geometry.

mathematical objects themselves. For example, in high school, students are introduced to tangent lines to circles. Rather than just telling students that a tangent line is perpendicular to the radius of the circle, using dynamic geometry tools, they can quickly explore this relationship and make conjectures about it. You might begin with a dynamic sketch of each that students can interact with and ask them what they notice and wonder about (Figure 1.3). Students might share that they notice that the line on the left intersects the circle in two points most of the time and the line on the right only has one point on the circle, or they might note that the angle between the line and the radius changes on the left and is constant on the right. All great observations prior to defining secant and tangent lines.

After providing some common language (e.g., a line that intersects a circle in two points is called a secant line, and a line that intersects a circle in exactly one point is called a tangent line), students can explore more as they are prompted to consider what happens as point C approaches and then coincides with point B and the relationship between the line and the radius when they coincide (see the images in Figure 1.4). Because they drag and measure, these properties are easily noticeable.

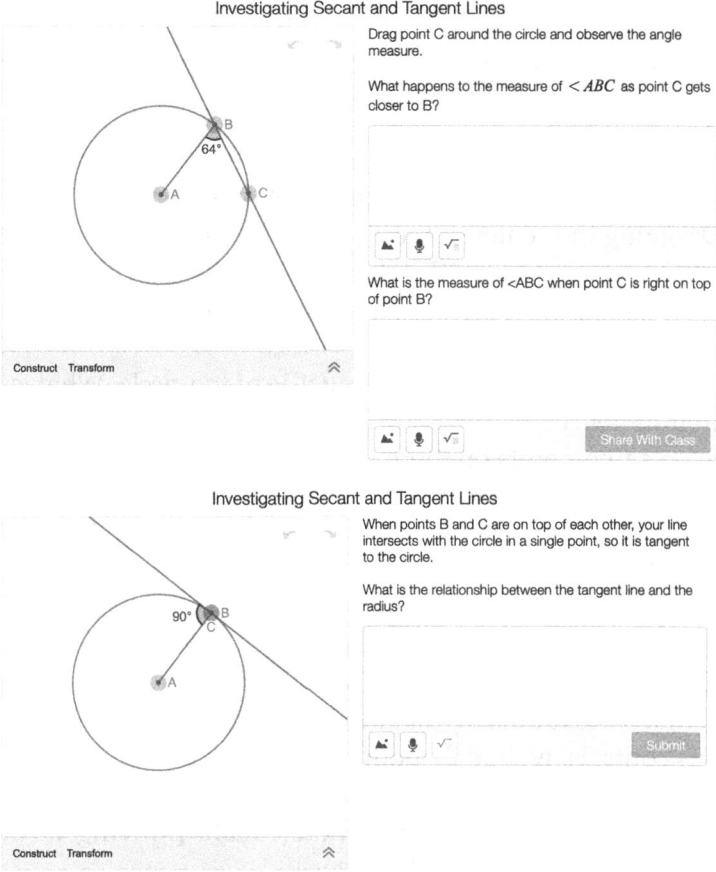

Figure 1.4 Further investigation of secant and tangent lines.

PAUSE AND CONSIDER

In the following vignette, you will see Isaiah and Sam investigating secant and tangent lines to determine the relationship between a tangent line and radius at the point of tangency. While you are reading/watching consider the following questions:

1. In what ways did the task + technology allow Isaiah and Sam to enter the problem?
2. How did the task + technology help Isaiah and Sam communicate their mathematical ideas?
3. How did the task + technology allow Isaiah and Sam to build an understanding of the relationship between a tangent line and radius at the point of tangency?

Vignette: Isaiah and Sam Investigate Tangent Lines to a Circle

Sam:	[Dragging the center of the circle around the screen.] All right, well that controls … okay. Well, the lines always touching a circle.	1 2
Isaiah:	Yeah. [Dragging the end point of the chord that is not the end point of the radius.]	3 4
Sam:	This one goes around like the outside of the circle. What is that? The circumference?	5 6
Isaiah:	There's two points on there.	7
Sam:	Yeah, but what is when you measure that?	8
Isaiah:	Circumference. [Dragging the endpoint of the radius that is on the circle.]	9 10
Sam:	Yeah, circumference and then this controls the size of it. [Dragging the end point of the chord again.]	11 12
Sam:	This controls the circumference.	13
Isaiah:	So that's the radius. [Dragging the center of the circle]	14
Sam:	This controls, wait, they both control size. Okay.	15
Isaiah:	[unclear]	16
Sam:	Is there another thing down here? Yeah. [Moving to the second sketch. Dragging the circle center.]	17 18
Sam:	Okay, this controls that movement and size.	19
Isaiah:	That one just has the radius. [Dragging the point of tangency]	20
Sam:	Oh, yeah.	21
Isaiah:	I bet the bottom one has a … Is there one on the bottom? [Moves back up on the page to type into the answer box.]	22 23
Sam:	So, the bottom one only has radius.	24
Isaiah:	Yeah, it only has radius. And this one has another point, which gives it an angle.	25 26
Sam:	And the top on has … what is it missing, though? Okay, cuz this one straight and this one's at an angle. So, it goes at an angle.	27 28
Isaiah:	Mhm.	29
Sam:	Wait, does this angle change? [Looking at the first sketch. Dragging the intersection point of the radius and secant line.]	30 31
Sam:	Right? Oh, I don't think okay. [Moves to the second sketch. Dragging the point of tangency.]	32 33
Sam:	So this angle stays the same?	34
Isaiah:	Yeah. Oh, cuz those are perpendicular.	35
Sam:	Yeah, and then this one changes. So I think that's okay.	36

Isaiah:	So the circle has an angle measure.	37
	[Students move to the next page of the activity. They are dragging	38
	point C on the circle.]	39
Sam:	Wait, what happened?	40
Isaiah:	[unclear] [Dragging C toward B and then away from B]	41
Sam:	No, but, it gets smaller, right? [Dragging C counterclockwise around	42
	the circle, ending with C coinciding with B.]	43
Isaiah:	It gets smaller as it gets closer to the radius.	44
Sam:	No, it gets smaller until it passes this point, right? That line right	45
	there, and then goes back up.	46
Isaiah:	So, the closer we get to the point B, the bigger it is, the closer it gets	47
	to [Dragging C clockwise around the circle.]	48
Sam:	No but at a it resets. Right? ["At a" refers to when they have dragged	49
	C so that the secant line coincides with the radius.]	50
Isaiah:	Once it's opposite from point B it's … so measure ABC as point C	51
	gets closer to B is right on top of point B. [Dragging C so that is	52
	moves slightly back and forth over point B. Then they move to the	53
	next page in the activity.]	54
Isaiah:	[reading quietly] Tangent line and the radius. Which one was the	55
	tangent line? This is the tangent line, right?	56
Sam:	I think so. [Dragging point B off of point C counterclockwise around	57
	the circle.]	58
Sam:	Yeah. Wasn't this this secant line?	59
Isaiah:	Because that's gonna be the radius. That's no, that would be [Using	60
	the cursor to point out the line he is referring to, line BC and then	61
	moving point B around the circle again.]	62
Sam:	I forgot.	63
Isaiah:	That they would call it the secant line.	64
Sam:	You can just go back and look. [They go back to the page of the	65
	activity where secant and tangent are defined.]	66
Sam:	Okay, that's the point of tangency this is a secant.	67
Isaiah:	Oh, I got it. [They return to the activity page they had been working	68
	on. Drags point C so it coincides with point B]	69
Sam:	So that would be tangent. [Dragging point C away from point B	70
	and dragging it around the circle.]	71
Isaiah:	Okay, cool. [*Reading the directions*] When points B and C are on top	72
	of each other, your line intersects with the circle in a single point, so	73
	it is tangent to the circle. What is the relationship between the tan-	74
	gent line and the radius? They're perpendicular. Pretty sure, right?	75
	[Drags point C so that it coincides with B and the measure shows 90	76
	degrees.]	77
Isaiah:	Cuz it's at 90 degrees.	78

> **PAUSE AND RECONSIDER**
> 1. In what ways did the task + technology allow Isaiah and Sam to enter the problem?
> 2. How did the task + technology help Isaiah and Sam communicate their mathematical ideas?
> 3. How did the task + technology allow Isaiah and Sam to build an understanding of the relationship between a tangent line and radius at the point of tangency?

What we love about this task is that it is not complicated, and it represents how many geometric definitions and theorems could be introduced – through exploration. In fact, the students could have constructed these images to explore themselves. However, the fact that the sketches are preconstructed allows the students to jump right in and start dragging objects to see what happens. Notice that as they begin dragging, they are just exploring what objects seem to be in the constructions and what happens when they drag each. It is not a focused exploration but one that allows students to get their bearings and to notice and wonder. They notice what controls the size of the circle, which segments appear to create angles, what type of angles, and which appear to move the entire figure.

If we read through just what the students say in the vignette, it is difficult to make sense of the students' thinking, but we can tell they are definitely talking with each other about the mathematical objects they are exploring. An important part of this communication is accomplished by showing each other what they noticed by dragging, observing what happens as they do so, and using gestures to point and recap what they observe. So while the vocabulary is new to the students, they are communicating clearly with each other using dragging to supplement their limited mathematical vocabulary.

The way Isaiah and Sam communicate with each other using the technology is key to their development of informal ideas about secant and tangent lines to circles. Through their exploration, they come to articulate the similarities and differences between secant and tangent lines with respect to the radius of a circle. They quickly decide that when they change the location of a secant line, it forms an angle with the radius that also changes. They also notice that the angle between the tangent line and the radius doesn't change. Moving on to the second page, where they are provided with measures as they drag point B to coincide with point C they conjecture that a tangent line not only has an angle that doesn't change at the point of tangency but also that the measure of that angle is 90 degrees. So, while they have not formalized this idea by proving it to be true, they have conjectured based on evidence that a tangent line is perpendicular to the radius at the point of tangency.

As a next step, we could ask Isaiah and Sam to use what they have conjectured to construct a line that is always tangent to a given circle. This requires using the properties they determined in their exploration and would provide further insight into their understanding of the concepts.

Example 2: Concept of Function

A common strategy to help students develop an understanding of the function concept is to use a function machine metaphor. A quick internet search will reveal many function machine applets. Often these applets follow the structure of a "guess my rule" machine (Figure 1.5). The user enters a number (input) and the machine applies the rule to determine and give the resulting number (output) to the user. However, in the next example, you will see students engaging with a function machine that was designed to contain no numerical or algebraic expressions and was built using a vending machine metaphor (Figure 1.6). This was done so that students would not focus on specific numerical values but instead would attend to the relationship between

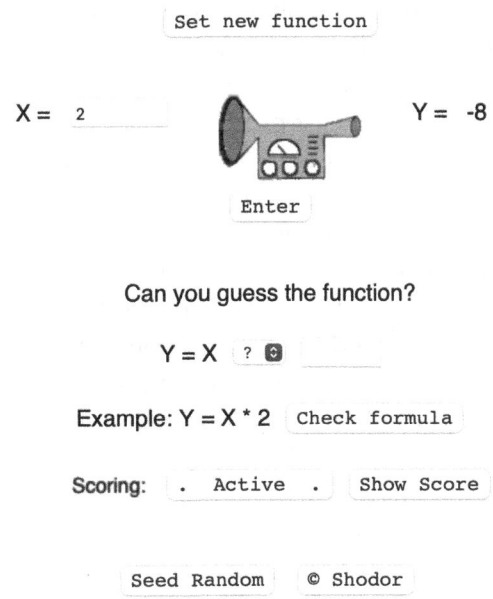

Figure 1.5 "Guess my rule" function machine applet by Shodor.

Which machine is a function?

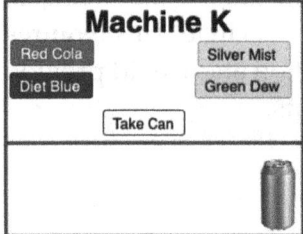 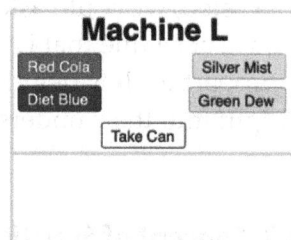

Don't forget to click Take Can each time.

Figure 1.6 Introduction to function applet.

the input and output. The idea was that by removing all numbers students would not form a reliance on the vertical line test procedure without developing a conceptual understanding of the function concept.

Link 1.3 "Guess My Rule" Function Machine Applet by Shodor

Link 1.4 Introduction to Function Applet

This GeoGebra book, Introduction to Function, has eight pages (Link 1.4). Each page consists of two vending machines, and each vending machine contains four buttons (Red Cola, Diet Blue, Silver Mist, and Green Dew). When a button is pressed on the machine, it produces none, one, or more of the four colored cans (red, blue, silver, green). On the first few pages when students are provided with two different machines, one is identified as a function and the other as not a function. The students are asked to explore the machines by clicking each button (input) to see the can(s) that the machine produces (output). They are not told how to click but to determine the difference between the two machines, they will need to click each button multiple times to compare the outputs. On later pages, students are shown two machines and asked which is a function and which is not a function. The goal of this task is for students to develop an informal definition of the concept of function through their interaction with the machines.

Next we're going to consider a vignette in which two seventh-grade students, Luz and Leah, are working on the Introduction to Function task. Neither had been introduced to the mathematical term *function* before this experience. Luz's and Leah's class was co-taught by a mathematics teacher and an inclusion support teacher. The whole class was working on this task in pairs, with both teachers monitoring the entire class. One of the teachers you will hear in the video spoke only English, so she used Google Translate on her phone to support the communication between Luz and Leah.

PAUSE AND CONSIDER

In the following vignette/video, you will see Luz and Leah investigating Machines K and L to determine which of these two machines is a function and why. While you are reading/watching consider the following questions:

1. In what ways did the task + technology allow Luz and Leah to enter the problem?
2. How did the task + technology help Luz and Leah communicate their mathematical ideas?
3. How did the task + technology allow Luz and Leah to build an informal understanding of the concept of function?

Link 1.5 Video Vignette: Luz and Leah Identifying Functions

Vignette: Luz and Leah Identifying Functions

Teacher:	[Student clicks on the Red Cola button; two red colas come out of	1
	the vending machine.] All right. Red is two reds. [Student clicks	2
	Diet Blue button; a blue can comes out.]	3
Teacher:	Azul, blue.	4
Teacher:	Blue. [Student clicks on the Red Cola button; two red colas come	5
	out of the vending machine.]	6
Teacher:	Okay.	7
Leah:	Wait, what is blue again?	8
Teacher:	Silver, blanco. [Student clicks on the Silver Mist button, and a silver	9
	can comes out of the vending machine.]	10
Teacher:	Silver. Green is green. [Student clicks on the green button, and a	11
	green can comes out.]	12
Luz:	And then …	13
Teacher:	Silver was silver, Leah. And green was green. [Student clicks on the	14
	Silver Mist button, and a silver can comes out of the vending	15
	machine. Student clicks on the green button, and a green can comes	16
	out.]	17
Teacher:	Okay.	18

Teacher:	Rojo. [Student clicks on the Red Cola button, two green cans come out of the vending machine.]	19 20
Luz:	What?	21
Teacher:	What did we say it was the first time? [Student clicks on the Red Cola button; two red cans come out of the vending machine.]	22 23
Leah:	Red. Red.	24
Luz:	No [Student clicks on the Red Cola button; two silver cans come out of the vending machine.]	25 26
Leah:	No [Student clicks on the Red Cola button; two blue cans come out of the vending machine.]	27 28
Luz:	Oh, now it's blue.	29
Teacher:	Por qué? Porque son diferentes colores.	30
Teacher:	Diferentes colors?	31
Teacher:	Leah, why is it not a function?	32
Leah:	Because it changes every time you click it.	33
Teacher:	Because it changes colors. Okay!	34
Leah:	Because it changes every time you click it.	35
Teacher:	Because it changes colors. Okay!	36
Leah:	So no function?	37
Teacher:	No function [Students move on to machine L. Student clicks on the Red Cola button multiple times; a green can comes out of the vending machine each time.]	38 39 40
Luz:	It's green. [Student clicks on the Blue button; a green can comes out of the vending machine.]	41 42
Leah:	Hit the green. [Student hits the silver button, and a green can comes out. Then they hit the green button, and a green can comes out.]	43 44
Luz:	Oh. Haha.	45
Leah:	It's all green? It's all green! [The student clicks each of the red, blue, silver, and green buttons again always getting a green can.]	46 47
Leah:	That's crazy!	48
Teacher:	What do you notice?	49
Luz:	Que notastes?	50
Luz:	Que siempre sale solamente el color verde	51
Teacher:	The only color it gives out is green. So function or no function?	52
Luz:	Yes, function.	53
Leah:	They're all green.	54
Leah:	Function.	55
Teacher:	Why? Por qué?	56
Leah:	Because they're all green.	57
Teacher:	Okay.	58

Luz:	Porque siempre sale el mismo color, y no sé, ugh, I don't know,	59
	porque …	60
Teacher:	So it always comes out the same, you don't know.	61
Luz:	Sí, funciona porque siempre sale el mismo color y no sale revuelto.	62
Teacher:	So it always comes out the same color and it's not scrambled, yeah.	63
Teacher:	It's always the same. Good.	64
Leah:	So it's a function?	65
Teacher:	Yep.	66

PAUSE AND RECONSIDER

1. In what ways did the task + technology allow Luz and Leah to enter the problem?
2. How did the task + technology help Luz and Leah communicate their mathematical ideas?
3. How did the task + technology allow Luz and Leah to build an informal understanding of the concept of function?

One of our favorite things about this task is that most students have had a previous experience engaging with a vending machine, or if not, there is likely one nearby in the school to go check out. Those previous experiences can easily translate to the task because students can simply click the buttons and observe what happens. Prior to this moment Luz and Leah had engaged with several pages of the GeoGebra book and had discovered the importance of trying each button more than once. In the video, you hopefully noticed that they were exploring the machines by trying each button multiple times, conveying their understanding that one cannot determine if the machine is a function or not by testing the button only once.

Luz and Leah communicated with each other and the teacher using not only their words but also the technology. In this vignette, the teacher was mediating the discussion between the students by stating the colors of the cans in both Spanish and English as Luz clicked the buttons on Machine K. It is important to note that when a silver can appeared, the teacher referred to it as "blanco, silver" because Luz referred to the silver cans as "blanco" (i.e., white) at the beginning of the activity and the teacher chose to continue with Luz's description and not correct her.

As Luz continued to click buttons, we see and hear both Luz's and Leah's surprise when the Red Cola button was clicked multiple times and different-colored sets of two cans showed up. They both immediately said "no" when the different set of two cans appeared. With Machine L, the teacher spoke less as the students communicated their ideas and engaged with the task. For example, Luz explained the machine as "solamente el color verde," and she used the mouse to circle Machine L to communicate which machine she was talking about. So even if the teacher or another student did not know what she said, she was communicating very clearly using the technology. Being able to use the cursor to point at and circle objects while using minimal words to communicate important ideas was vital to their work together.

Not only were Luz and Leah communicating with and through the task + technology, but it is also evident that they were developing powerful informal ideas about the concept of function. For example, at the end of the clip, Leah explained that Machine L is a function because "they are all green"; Luz added to that idea by explaining, "Siempre sale el mismo color y no sale revuelto" (Google Translate recognized this for the teacher as "It always comes out the same color, and it's not scrambled"). The students engaged with the task and with each other without the use of formal mathematics language or even being fluent in the same language. The way the technology was used in this task positioned them as powerful math explorers. They were not told what a function was but instead were trusted with defining it themselves before returning to a whole-class discussion to formally define function.

Concluding Thoughts

Whether comparing measures of center, tangent lines to a circle, or the concept of function, the use of technology in each of the tasks described positioned students to ask questions. Questions like "What happens if?" "Is this always true?" and "Why?" They were true mathematical explorers. In each there was a way for students to act on an object and see what happens as a result of their actions. The dynamic nature of the technology used in the task facilitated their exploration.

The reason we even think about incorporating technology into math tasks is because of the ways it supports students as explorers. Technology can help to make complex mathematical ideas more accessible because students are invited to jump in and explore by acting on an object, regardless of the prior experiences they are bringing to the table. It is through exploration that technology has the potential to mediate mathematical discussions, support student sensemaking and reasoning, and the development of one's identity as a knower and doer of mathematics.

CHAPTER TAKEAWAYS

Think of your students as math explorers! When incorporating technology-enhanced tasks, look for the ways it positions all students to do the following:

- Access and explore the math
- Communicate their mathematical ideas
- Develop informal and powerful ways of thinking mathematically

---

Questions to Discuss With Your Colleagues

1. In the introduction to this chapter, it was suggested that we think of students as math explorers. How might doing that influence the tasks you select for your students to engage with? Can you think of specific characteristics you might look for in those tasks you select?
2. In this chapter, you read about the ways we use technology to perpetuate inequities in classrooms. Did any of those descriptions of inequity resonate with you? Why or why not?
3. If you have used any technology-enhanced math tasks recently (or have seen others use them), in what ways did they position students as math explorers? If they did not position students as math explorers, how could they be adapted so that they do?

 You can find links to all the technology-enhanced tasks and supplementary video throughout the book at https://www.tlmtresearch.com/teachingmathtechbook.

2

Math Specific Technology Tools

Now that you have seen how technology can position students as math explorers, you may be wondering what different technologies are out there that can be used to support exploration. If students are going to be able to explore, it is important that they can act on mathematical objects and the technology responds to those actions in mathematically defined ways. These kinds of technologies allow students to be explorers by considering multiple (sometimes infinite) examples, diverse or extreme examples, and looking systematically at cases they create (Belnap & Parrot, 2020). There are seven types of math action technologies that are typically used to support students as explorers in secondary mathematics, including dynamic graphing applications, dynamic geometry applications, data analysis applications, computer algebra systems (CASs), spreadsheets, virtual manipulatives, and interactive applets (see Table 2.1).

All these technologies can be used on computers, tablets, and cell phones, making them widely accessible. But most important, they all allow students to explore mathematical ideas and make and test conjectures about mathematical patterns they notice. Students need to be familiar with a wide variety of tools and recognize when particular tools might be helpful and know how to use them to explore and deepen their understanding of concepts (National Governors Association Center for Best Practices & Council of Chief State School Officers, 2010). To that end, the purpose of this chapter is to describe each of the seven types of math action technologies in detail and provide some examples of the ways they support exploration.

DOI: 10.4324/9781003302285-3

Table 2.1 Types of Math Action Technologies

Type	Description
Dynamic graphing applications	Graphing applications allow students to observe and explore multiple representations of functions and data by generating linked tables, graphs, and symbolic representations. These include graphing calculators.
Dynamic geometry applications	Dynamic geometry software allows for an exploration of geometric relationships in coordinate, transformational, and synthetic contexts. Students can make and explore conjectures through dragging geometric objects and attending to invariances. Such explorations can provide insight into the existence of relationships and why they hold true, an important step to generating formal proofs.
Data analysis applications	Data analysis tools support the visualization of large data sets with linked representations and tools for simulating observable phenomena. These tools provide opportunities to explore "what if" questions that are invaluable to the study of probability and statistics.
Computer algebra systems (CASs)/ Dynamic algebra notation systems (DANSs)	CASs can operate on algebraic statements. This allows for insight into the structure of algebraic functions and expressions and is especially powerful for highlighting patterns of equivalence, such as factoring quadratic equations. Similarly, DANSs allow students to act directly on symbolic representations.
Spreadsheets	Spreadsheet applications quickly display the results of repeated calculations and can generate tables of values linked to a variety of graphical representations. Displaying repeated calculations allows for insights into structure and relationships among variables.

Table 2.1 (Continued)

Type	Description
Virtual manipulatives	Virtual manipulatives, created to parallel their physical counterparts, allow for dynamic observation and the creation of flexible visual representations of mathematical ideas. The virtual aspect often affords more flexibility and the ability to save drafts that is not possible with physical versions.
Interactive applets	Small exploratory computer programs that allow students to dynamically interact with objects and observe them in mathematical ways (sometimes referred to as microworlds). When carefully selected to align with a mathematical goal, applets provide a way for students to quickly and easily explore mathematical relationships.

Types of Math Action Technologies

PAUSE AND CONSIDER

1. Which of the types of technologies in Table 2.1 have you used in the past?
2. In what ways have you used them to position students as math explorers as they build their understanding of important secondary math concepts?
3. Which are you most excited to learn more about? Why?

Link 2.1 List of Specific Math Action Technologies in Each Category

Dynamic Graphing Applications

Probably the most commonly used type of technology in secondary classrooms are dynamic graphing applications (e.g., TI 84 handheld graphing calculator, Desmos graphing calculator). These technologies are characterized by the ability to create linked tables, graphs, and symbolic representations. Since the representations are dynamically linked, students can act on one object (e.g., change a term in an equation) and the others (e.g., table and graph) will change accordingly. This allows students to reason about algebraic structure

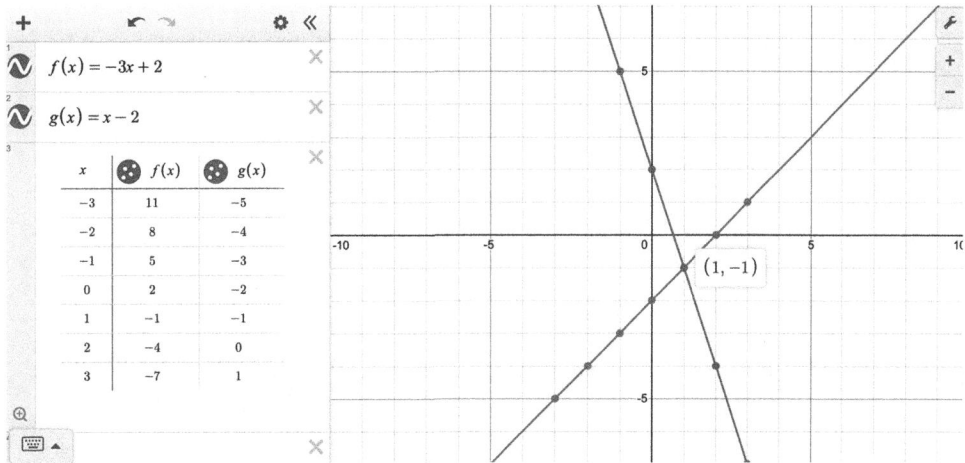

Figure 2.1 Tabular and graphical representations of a solution to an equation.

from different perspectives, which is especially useful in supporting students reasoning about solutions to equations or systems of equations.

For example, imagine students exploring what it means to be a solution to an equation. When thinking about solving equations as identifying all possible values that make the statement true, tables and graphs can help by creating an image of what it means to be "true". Consider the equation $-3x + 2 = x - 2$. To identify the values that make the statement true (i.e., the solution[s]), students can create a table linked to a graph as seen in Figure 2.1. In the table, one can see that when $x = 1$, both the left and right side of the equation equal -1 (i.e., $-1 = -1$), making the statement true. They can then see this represented in the graph where the two functions intersect. The same sort of exploration can also be done with inequalities.

Dynamic graphing applications are also very helpful for exploring the key characteristics of function families. Consider the dynamic graph of $f(x) = a \sin(bx) + k$ shown in Figure 2.2. The sliders are dynamic. This means that as the point on the slider is dragged, the value changes, is substituted into the function, and the graph changes accordingly. Students can drag the sliders to explore the relationships among the parameters of the function (a, b, and k) and the amplitude, midline, and period of the graph.

In addition to having dynamic tools to explore equations and functions, dynamic graphing technology can also be used to explore statistical ideas. Not only do they have built-in functions for determining various statistical measures when a list of data is entered (e.g., mean, median, mode, standard deviation, interquartile range, maximum and minimum), but as was shown in Chapter 1, the dynamic graphing capabilities can also be used to

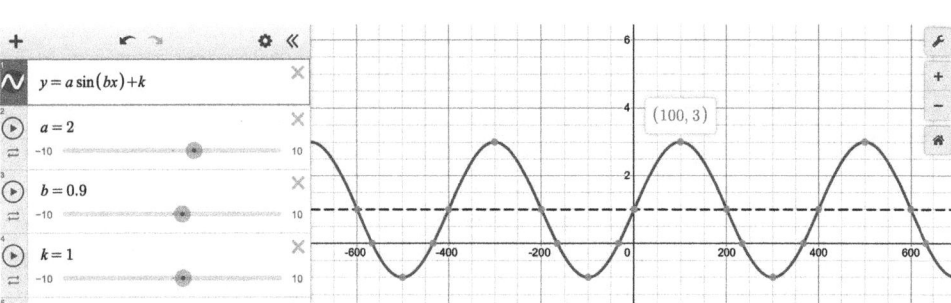

Figure 2.2 Graph of $y = a \sin(bx) + k$ with sliders for a, b, and k.

create dynamic dot plots, box plots, and histograms that allow students to drag individual data points and explore the effect on the shape, center, and spread.

Dynamic Geometry Applications

Dynamic geometry applications (e.g., GeoGebra, The Geometer's Sketchpad) are characterized by the ability to construct geometric objects (e.g., points, lines, circles) using electronic versions of the historical tools of the geometric trade – compass and straightedge – and being able to drag and maintain the mathematical properties of the constructed objects. For example, as we saw in Chapter 1, students can use the circle tool to construct a circle with center A and put a point B on the circle. They can use the segment tool to create segment AB, a radius of circle A. Then they can add another point, C, on the circle and use the line tool to create secant $\overline{BC}$. Next, they can measure $\angle ABC$. As students drag C, the measure of $\angle ABC$ will change accordingly. In fact, students can drag point C toward B, and the measure of the angle will approach 90 degrees. With point C directly on top of B, the measure of $\angle ABC$ will be 90 degrees. Such an exploration provides the opportunity to build an understanding of the difference between a secant line and a tangent line as well as the fact that a tangent line is perpendicular to the radius at the point of tangency (see Figure 2.3).

Dragging a constructed object, like the earlier example, results in creating infinite examples of the objects with the geometric properties of the construction remaining intact. This allows students to explore conjectures through dragging geometric objects and attending to invariances (i.e., noticing what changes and what doesn't as you drag). This allows for explorations that can provide insight into the existence of relationships and why they hold true, an important step toward generating formal proofs.

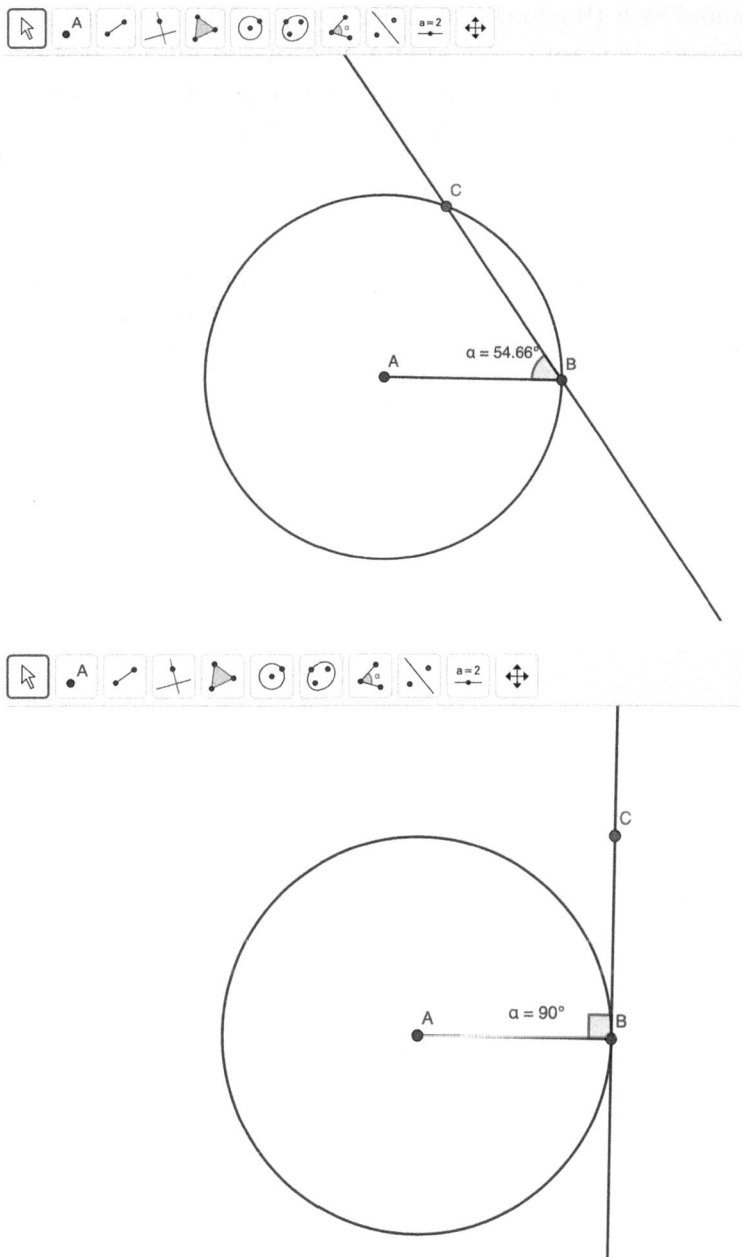

Figure 2.3 Dragging a point on a secant line so it becomes a tangent line.

For example, consider the theorem, "if a point lies on the perpendicular bisector of a segment, then it is equidistant from the end points of the segment". One way of introducing this theorem is by posing a situation in which it would be useful. A common context used for this is *the building a new playground* task (Boston et al., 2017, p. 91). This problem states that a local commission is wanting to build a new playground that is equidistant from two elementary schools. Using a dynamic geometry platform students could place two points in the plane to represent the two elementary schools and then be asked to determine at least three different locations for the playground using the tools available to them.

To construct their solutions students will likely start by finding the midpoint of the segment between the two points as they know this location will be equidistant to both end points (i.e., schools). To find additional locations, they might do the following (see Figure 2.4 for examples):

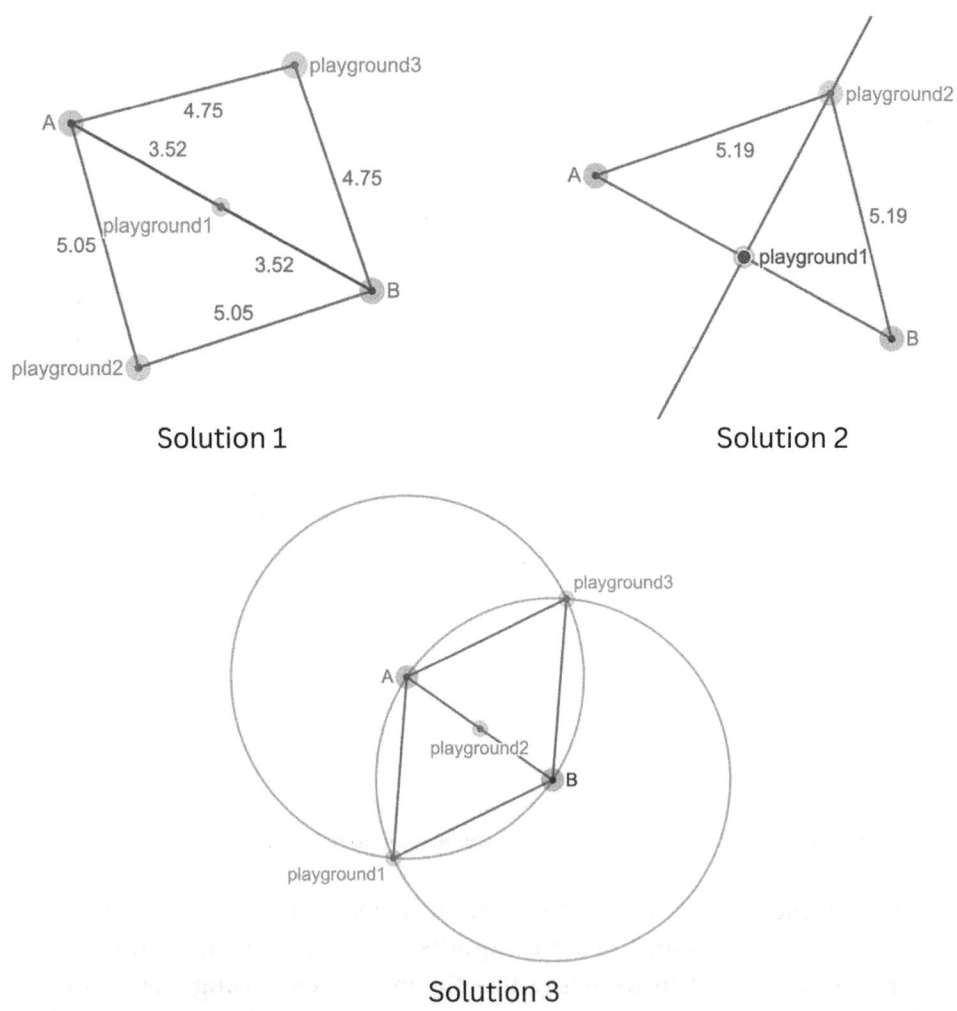

Figure 2.4 Sample solution strategies.

1. Create a point C and segments *AC* and *BC*. Then after adding a measure for AC and BC, drag point C until the segment lengths are congruent.
2. Create segment *AD* and line *GH*. Then knowing that a reflection preserves length, use the reflect tool to reflect segment *AD* over line *GH* and drag the line of reflection until finding a place where the reflected segment endpoints lay on D and B.
3. Create two intersecting circles, both with radius *AB*. Then using the fact that radii of a given circle are congruent and circles with congruent radii are congruent to each other, find the intersection(s) of the two circles.

Students should be prompted to make sure their solution would work no matter where the schools were placed (i.e., if they drag a point that represents a school, does their playground location still work?). Once they have determined at least three locations, ask what they notice. Next ask the students to use what they noticed to find the set of all possible locations for the playground. This leads to a nice conjecture; it seems as if a point lies on the perpendicular bisector of a segment, it is equidistant from the end points of the segment (our theorem). Furthermore, the exploration itself provided some very important insight to the key ideas of the proof of the theorem (Raman & Weber, 2006). Regardless of the construction method they chose, students will have created a series of right triangles that appear to be congruent – they would have talked about the hypotenuse and the leg being congruent to explain why their locations are solutions. Recognizing this, and that it means the two triangles are congruent, are key ideas of the proof that once identified will help students get started on the details of the formal proof.

Data Exploration and Analysis Applications

Data exploration and analysis applications (e.g., CODAP, Tuva, Tinkerplots, Fathom) are characterized by the ability to create visualizations of large data sets with linked representations. This allows for the exploration of multiple numerical and categorical variables through the creation of dot plots, box plots, and histograms, as well as statistical measures. The statistical measures can be displayed on the graphical representations to support sensemaking about them.

For example, using data on a variety of common cereals found in grocery stores, students can explore questions like "Which shelf should you select your cereal from if you want to make sure it is healthier?" Students can explore the question by dragging and dropping various variables to create data representations and creating statistics toward crafting an argument that one shelf or another contains healthier cereal options.

For example, using CODAP, a student might decide to look at "sugars" and drag and drop that variable from the table on the x-axis. They can then split the data by dragging "shelf" and dropping it on the y-axis. Then they can add the measures of center (indicated by the vertical lines) and box plots to support their descriptions and comparisons of the data sets (see Figure 2.5). The table and graph are dynamically linked. Clicking on a row in the table will highlight the observation of that row in the graph. Similarly, clicking on an observation in the graph will highlight the associated row in the table. Hiding, deleting, or adding data in the table will result in dynamic changes in the graph to match the changes. Importantly, the ability to drag and drop variables allows for the exploration of multiple categorical variables. In addition, the speed at which these representations can be created promotes quick the exploration of which variables to investigate more closely. The same drag-and-drop actions allow for exploration of association among continuous variables as well.

Often data exploration and analysis applications also include tools for simulating observable phenomena. For example, consider the Illustrative Mathematics *Valentine Marbles* task in which students investigate a Valentine's Day contest held by a hotel in which guests are invited to estimate the percentage of red marbles in a huge clear jar containing both red and white marbles. In this task, the hotel staff know there are 11,000 marbles in the jar (3696 are red, 7304 are white). The guests are allowed to take a random sample of 16 marbles from the jar in order to come up with their estimate. Rather than just showing an image of a sampling distribution, we can simulate the process of 100 guests each reaching in and randomly selecting 16 marbles. In fact, we can do this multiple times so that students can see that over time this process results in a sampling distribution that approximates the proportion of red marbles. We can discuss why this result makes sense and then how we might use this information to make inferences about the situation (Figure 2.6).

Link 2.2 The Valentine Marbles Task

Link 2.3 CODAP Simulation of the Valentine Marble Task

Computer Algebra Systems and Dynamic Algebra Notation Systems

CASs (e.g., GeoGebra CAS, TI-Nspire plus CAS) are characterized by the ability to operate on algebraic statements. For example, entering "factor $(x^2 + 5x + 6)$" into a CAS results in $(x + 2)(x + 3)$. CAS commands include, but are not limited to, *expand, solve, integral, derivative, GCD, limit,* and *binomialdist*. Being able to act on algebraic statements allows for insight into the structure of algebraic functions and expressions, and is especially powerful

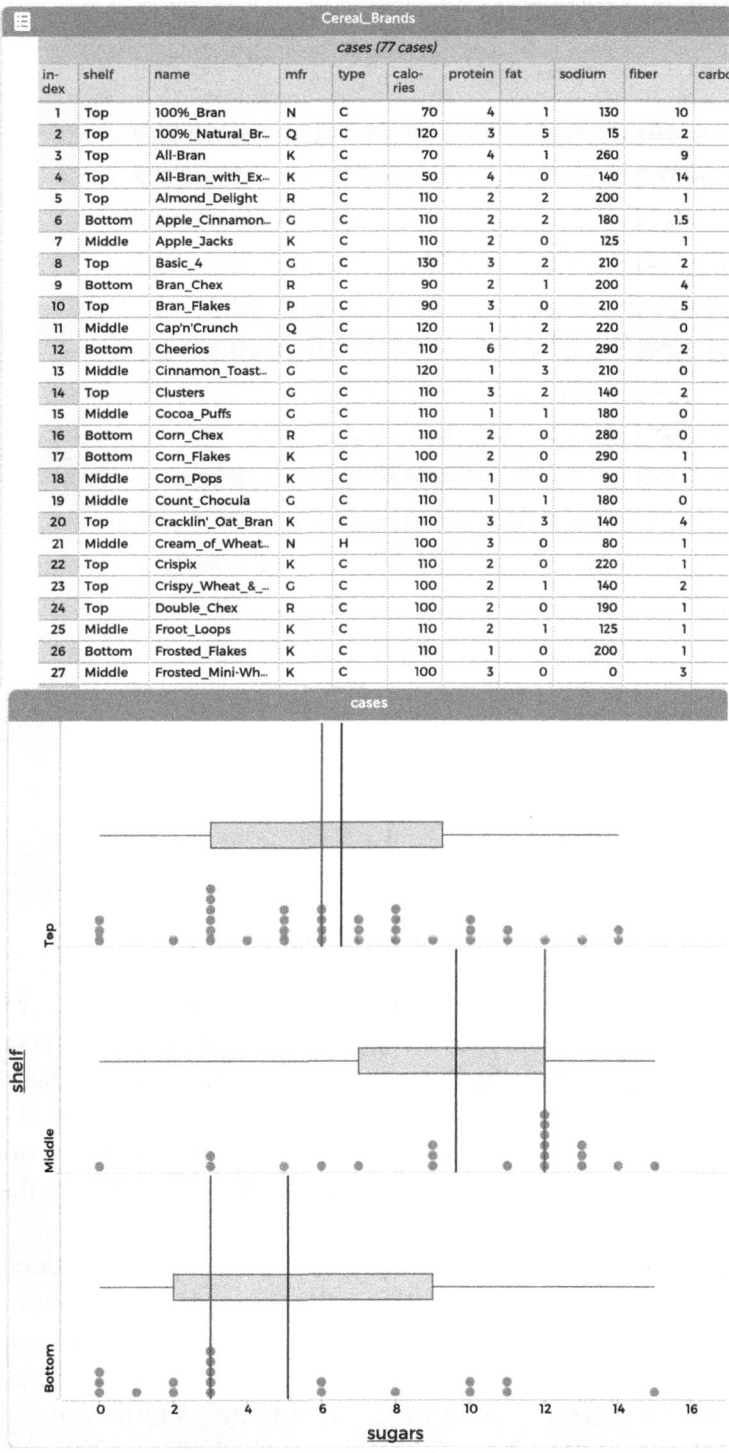

Figure 2.5 Comparing data sets using CODAP.

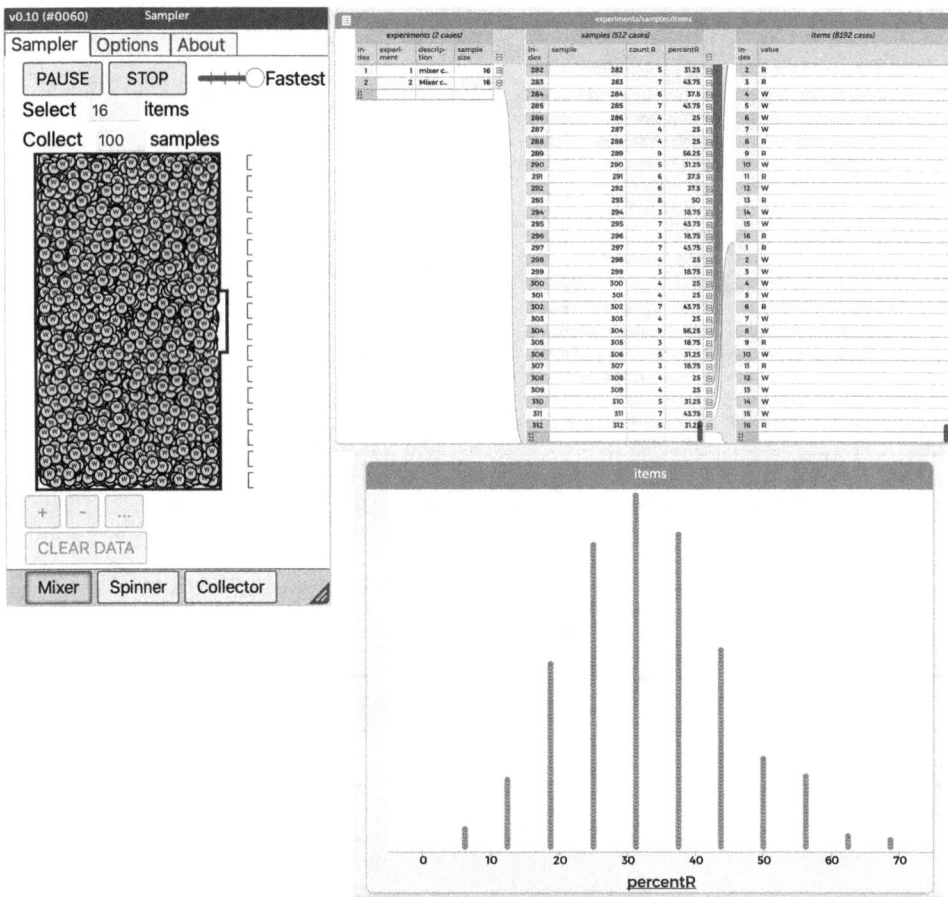

Figure 2.6 Simulating the valentines marble task with the CODAP sampler.

for highlighting patterns of equivalence, such as factoring quadratic equations. For example, imagine providing a carefully selected sequence of cubic expressions that are each either a sum or a difference of two cubes. Students can use a CAS to factor them and then describe the patterns they notice in the structure of both the original expressions and their factored forms (Figure 2.7). Based on this exploration, students can determine the rule for factoring a sum or difference of two cubes (Figure 2.8).

Attending to patterns of equivalence can also support students as they build their fluency in solving equations. For example, rather than using the *solve* command in a CAS, students can use their knowledge of properties of equality to solve equations without worrying about a computational error. In fact, computational errors would make identifying patterns very difficult. In some CAS platforms (the TI-Nspire CAS is shown in Figure 2.9), substituting potential solutions back into the original equation (i.e., checking our work)

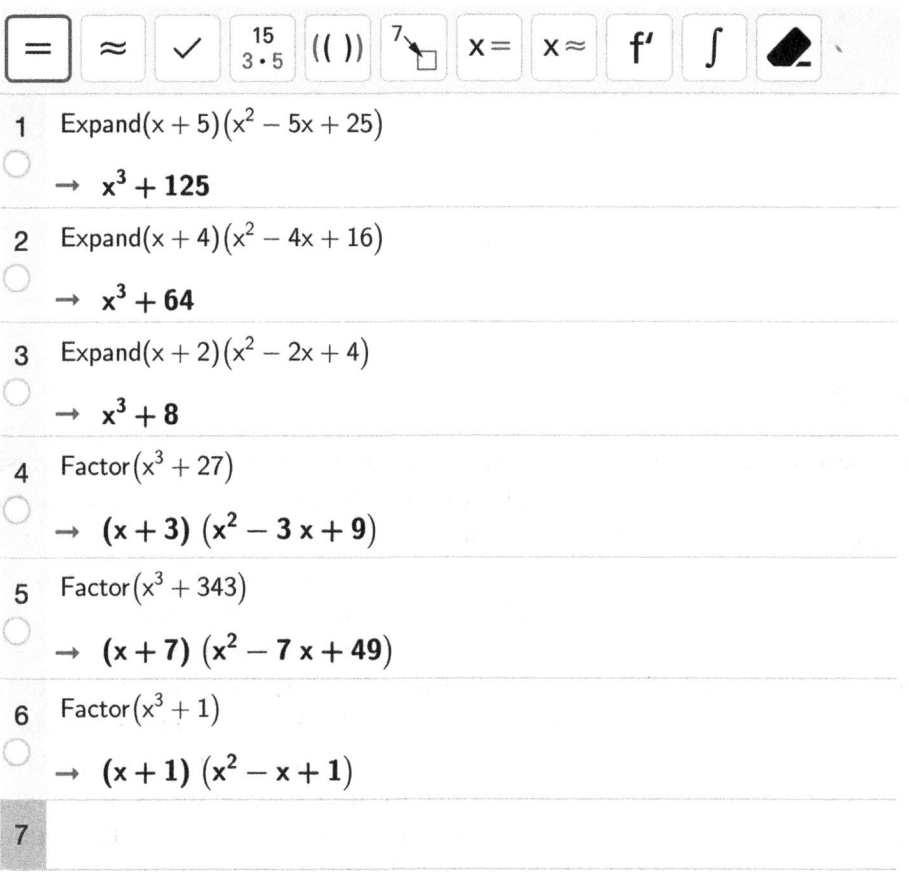

Figure 2.7 Using a CAS to examine structure when factoring a sum of cubes.

Look at your completed table. What do you notice? What do you wonder?
Please write down at least 4 things that you notice.

1) I notice that when you take out the second term in the second parentheses your product will be the expanded form

2) I WONDER HOW TO CONVERT IT FROM EXPANDED TO FACTOR FORM W/O DOING A BUNCH OF CRAZY MATH

1

$$(x+a)(x^2-ax+a^2) = x^3+a^3$$

Figure 2.8 Student response when asked to generalize after exploring using a CAS.

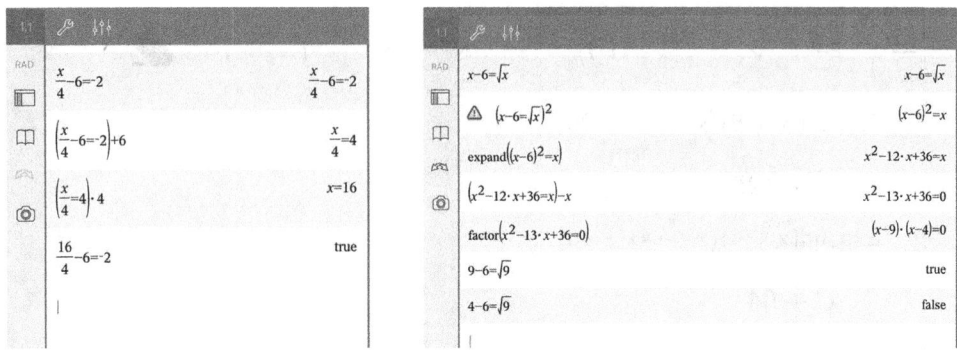

Figure 2.9 Using a CAS to solve and check an equation.

will result in statements that are either "true" or "false", further emphasizing that solving is determining values that make the equation statement true (Figure 2.9). Notice that in the second example, there is a warning that the "operation might introduce false solutions" (Figure 2.9). This warning provides an opportunity to discuss why the false solution occurred.

Dynamic algebra notation systems (DANSs), which are similar to CASs, are becoming increasingly popular as they allow students to act directly on symbolic notation. A commonly used free DANS technology is Graspable Math. In the example shown in Figure 2.10, the literal equation $Ax + By = C$ was given. Students can use their mouse to drag specific terms of the equation and perform operations. The second line is the result of the student dividing by A by dragging A to the right side of the equation. The technology responded by dividing all terms by A. To perform operations students can just click on the operation symbol or drag the terms to be operated on toward each other, and if it is possible to perform, the result will be shown. However, if it is not possible, then the symbol will shake, indicating the operation cannot be performed. When used appropriately, DANSs can support the development of procedural fluency.

Spreadsheets

Spreadsheet applications (e.g., Excel, Google Sheets) quickly display the results of repeated calculations and can generate tables of values linked to a variety of graphical representations. Displaying repeated calculations allows for insights into structure and relationships among variables (e.g., dependent vs. independent variables). For example, consider the classic Tortoise and the Hare task (modified here to include an alligator; McCulloch et al., 2015).

Turtle, Rabbit, and Alligator are competing in a 100-meter dash. Rabbit is excited, he knows he is the fastest of them all. He's bouncing around

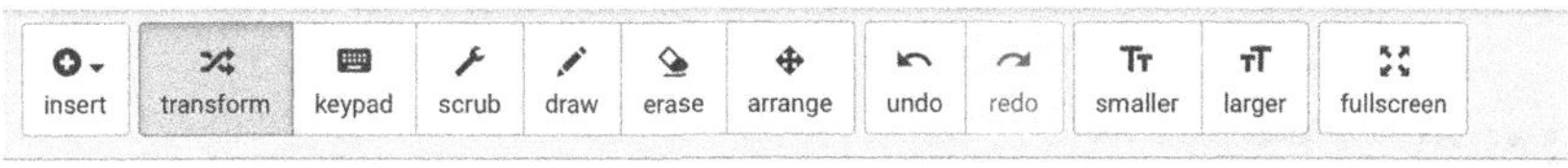

$$Ax + By = C$$

$$x + \frac{By}{A} = \frac{C}{A}$$

$$\frac{By}{A} = \frac{C}{A} - x$$

$$y = \frac{C - Ax}{B}$$

$$y = \frac{C}{B} - \frac{Ax}{B}$$

Figure 2.10 Using DANSs to solve a literal equation.

at the starting line taunting Turtle. "Turtle, you know I'm going to win, so why don't you just go ahead and start". Alligator looked at Turtle's shell and figured that was a lot to carry around, so he agreed with Rabbit's suggestion. They decided to give Turtle a 60-meter head start. So the race referee had Turtle line up 60 meters ahead of Rabbit and Alligator when he announced, "Take your mark … get set … go!" Rabbit took off sprinting and just kept getting faster, starting at a rate of 1 meter/sec and going 1 meter/sec faster each second of the race until he got to the finish line. Turtle took off at a nice steady walk and walked at a rate of 3 meters/sec for the duration of the race. When the starting signal sounded, Alligator was caught off guard. He stood still and saw that Rabbit had taken off quickly. At 5 seconds he finally started running. He began at a pace of 1.5 meters/sec and continued to speed up, going 1.5 times faster for every second of the race. The story obviously ends with the winner of the race, but who is it?

When building the three animals' race using the functions in a spreadsheet, students must figure out how to represent how each animal's distance from the starting line changes over time which requires paying particular attention to the rate of change for each. Using the spreadsheet formulas to carry out the

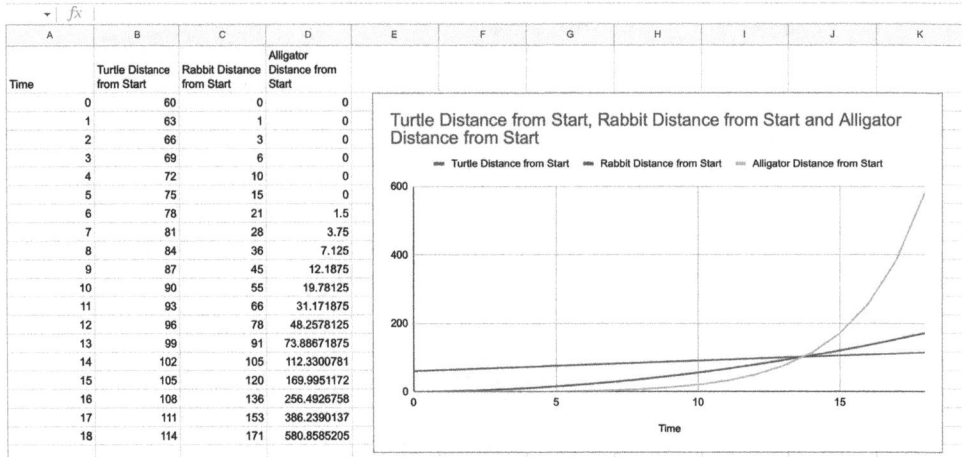

Time	Turtle Distance from Start	Rabbit Distance from Start	Alligator Distance from Start
0	60	0	0
1	63	1	0
2	66	3	0
3	69	6	0
4	72	10	0
5	75	15	0
6	78	21	1.5
7	81	28	3.75
8	84	36	7.125
9	87	45	12.1875
10	90	55	19.78125
11	93	66	31.171875
12	96	78	48.2578125
13	99	91	73.88671875
14	102	105	112.3300781
15	105	120	169.9951172
16	108	136	256.4926758
17	111	153	386.2390137
18	114	171	580.8585205

Figure 2.11 Using spreadsheets to examine structure and created linked representations.

recursive pattern supports students' sensemaking about the different ways in which each of the animals' rates changes, drawing clear distinctions between constant rates of change, constantly additive rates of change, and constantly multiplicative rates of change. Creating a linked graph of the data further emphasizes how the animals' relative speed and distance from the start compare to each other (Figure 2.11).

Given that spreadsheets allow for coding recursive situations, like the *Three Animals Race*, they can be very useful for exploring arithmetic and geometric sequences and series. See the spreadsheet image shown in Figure 2.12, where students have defined a starting value, a multiplier, and an add-on. They then use the spreadsheet functions to build the sequence and the series. Since the spreadsheet is built using functions, changing the multiplier (for example) dynamically changes the sequence and the series. This sort of activity could be used to explore the conditions under which a series converges or diverges.

Virtual Manipulatives

Virtual manipulatives, created to parallel their physical counterparts, allow for dynamic observation and the creation of flexible visual representations of mathematical ideas. Such manipulatives include, but are not limited to, pattern tiles, tangrams, three-dimensional (3D) solids, and algebra tiles. Virtual manipulatives can be very helpful in supporting students build sensemaking of abstract ideas with concrete objects. While not all schools have physical manipulatives, virtual options are freely available online (e.g., Polypad, Toytheater, NCTM Illuminations). For example, when teaching probability, it is common to use dice, cards, or spinners to simulate situations. Each of which is available virtually. The virtual aspect of these manipulatives often affords

C13 ▾ fx =C12+B13

	A	B	C
1	starting number	1	
2	multiplier	0.5	
3	add on	0	
4			
5	term number	sequence	series
6	0	1	1
7	1	0.5	1.5
8	2	0.25	1.75
9	3	0.125	1.875
10	4	0.0625	1.9375
11	5	0.03125	1.96875
12	6	0.015625	1.984375
13	7	0.0078125	1.9921875
14	8	0.00390625	1.99609375
15	9	0.001953125	1.998046875
16	10	0.0009765625	1.999023438
17	11	0.00048828125	1.999511719
18	12	0.000244140625	1.999755859
19	13	0.000122070312	1.99987793
20	14	0.000061035156	1.999938965
21	15	0.000030517578	1.999969482
22	16	0.000015258789	1.999984741
23	17	0.000007629394	1.999992371
24	18	0.000003814697	1.999996185

B13 ▾ fx =B12*B2+B3

	A	B	C
1	starting number	1	
2	multiplier	2	
3	add on	0	
4			
5	term number	sequence	series
6	0	1	1
7	1	2	3
8	2	4	7
9	3	8	15
10	4	16	31
11	5	32	63
12	6	64	127
13	7	128	255
14	8	256	511
15	9	512	1023
16	10	1024	2047
17	11	2048	4095
18	12	4096	8191
19	13	8192	16383
20	14	16384	32767
21	15	32768	65535
22	16	65536	131071
23	17	131072	262143
24	18	262144	524287

Figure 2.12 Using spreadsheets to explore arithmetic and geometric sequences and series.

more flexibility than the physical forms. For example, consider spinners that allow for changing the number of sections or even the size of those sections like the example from NCTM Illuminations in Link 2.4.

Link 2.4 Adjustable Spinner

One of the most commonly used manipulatives in secondary math is algebra tiles. Algebra tiles can be used to represent polynomial operations ranging from combining like terms to factoring. For example, using algebra tiles to represent $(2x^2 + 3x + 4) + (-5x - 1)$ students would first lay out the tiles to represent $2x^2 + 3x + 4$ and $-5x - 1$ (Figure 2.13 on the left). Then combining like terms, it is possible to see that $x + -x$ results in 0, as does $1 + (-1)$ (i.e., they are zero pairs). Thus, the simplified expression is $2x^2 - 2x + 3$ (Figure 2.13 on the right).

Virtual algebra tiles are invaluable in secondary math as they can be used to build on students' understanding of multiplication and division using area models to extend to the distributive property, factoring, and even completing the square (shown here are the virtual algebra tiles in Polypad, Link 2.4). Consider the example shown in Figure 2.14 in which a student used algebra tiles to complete the square for the expression $x^2 - 8x + 9$. To complete the square the students' saw they needed 7 ones, but along with them included

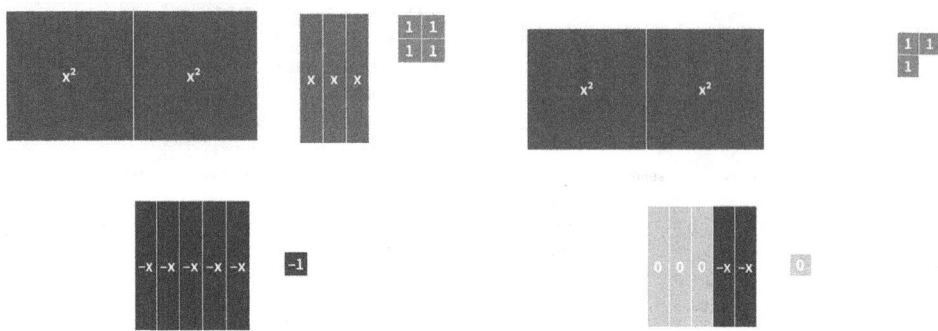

Figure 2.13 Representing $(2x^2 + 3x + 4) + (-5x - 1)$ with algebra tiles.

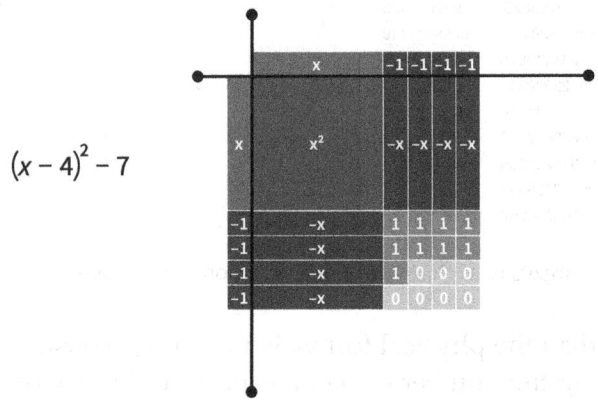

$(x - 4)^2 - 7$

Figure 2.14 Using algebra tiles to complete the square.

7 negative ones as well so as not to change the value of the expression. These show up as 0 when paired together, indicating the value of the expression has not changed. Thus, the completed square has length and width $(x - 4)$, and the 7 negative ones are included in the expression; it remains equivalent to the original expression (i.e., $(x - 4)^2 - 7$).

Link 2.5 Polypad Virtual Algebra Tiles

Interactive Applets

Interactive applets are small computer programs that students can interact with to explore mathematics. Unlike large software that allows for a variety of mathematics to be explored, interactive applets are designed to engage students in a single mathematical topic. Teachers can build their own applets using technologies like GeoGebra (e.g., the Vending Machine applet shown in Chapter 1), but there are also many collections that are premade by educational

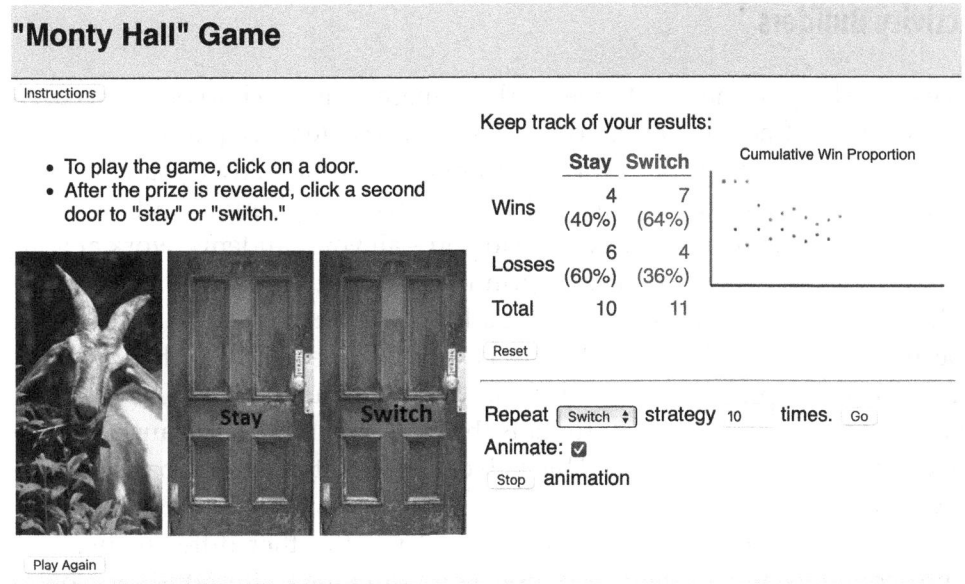

Figure 2.15 Investigating the "Monty Hall" game.

designers for teachers to use (e.g., NCTM's Illuminations Interactivities, Rossman/Chance Applets, PhET Interactive Simulations). For example, among NCTM's Illuminations *Interactivities* collection is an applet that supports the exploration of properties of geometric solids (Link 2.6), Rossman/Chance has an applet that simulates the Monty Hall Game (Figure 2.15 and Link 2.7), and PhET has one that supports student understanding of residuals and the least squares regression line (Link 2.8).

Link 2.6 Investigating Geometric Solids

Link 2.7 "Monty Hall" Game

Link 2.8 Least Squares Regression Interactive Applet

There are also interactive applets for common demonstrations used in secondary mathematics. Consider the conic sections. There exist physical 3D models of cones that have been sliced in different ways to illustrate the conic sections, but if they are not available in schools already, they are expensive to purchase. However, there are nice online models. For example, this virtual model by Interactive Math (Link 2.9) allows for the location of the intersecting plane to be changed to see the resulting cross section.

Link 2.9 Conic Sections Model

Activity Builders `

One of the biggest changes to the world of math action technologies has been the addition of activity builders with specialized teacher platforms. These spaces allow you to package task prompts that accompany dynamic representations in such a way that student responses to prompts and creations with the math action tools are visible to you – all your students' work at your fingertips at any time! As of this printing Desmos, GeoGebra, and Graspable Math all have their own activity builders with teacher views. Non-math-specific activity builders, like PearDeck can also be used for this purpose. Many of these platforms (e.g., Desmos Activity Builder, GeoGebra Classroom) also have ways to pace students through an activity and pause them so they can't use the technology. (This will cause them to loudly groan and turn to you to see what's up!)

Activity builders allow you to easily string together different types of representations and student responses (e.g., drawings, short answers, paragraph explanations, graphs, constructions, multiple choice responses) to create a storyline for your task or lesson. We think of activity builders as the conveyance and collaboration spaces in which math action technologies can be embedded. That means that they can be used in really powerful ways – but that also means that if you aren't careful, they can also simply end up being used as digital worksheets. We will be referencing these activity builders throughout this book, with a focus on how to use them with embedded math action technology.

Concluding Thoughts

Our goal in this chapter was to provide a brief overview of the most common types of math action technologies used in secondary math. In doing so, we included some examples of how each could be used to support students' math exploration. Something additional to consider is that sometimes it is helpful to combine technologies. For example, GeoGebra includes both dynamic geometry and dynamic graphing tools which allows for using both together – a very powerful option for making sense of trigonometric functions and their graphs (see Figure 2.16, a GeoGebra applet created by Daniel Mentrard). It is not uncommon to use spreadsheets and data exploration software together. Large data, whether collected through surveys or found in existing data collections, can often be easily downloaded into a spreadsheet. After cleaning the data, one might do some initial analysis using spreadsheet

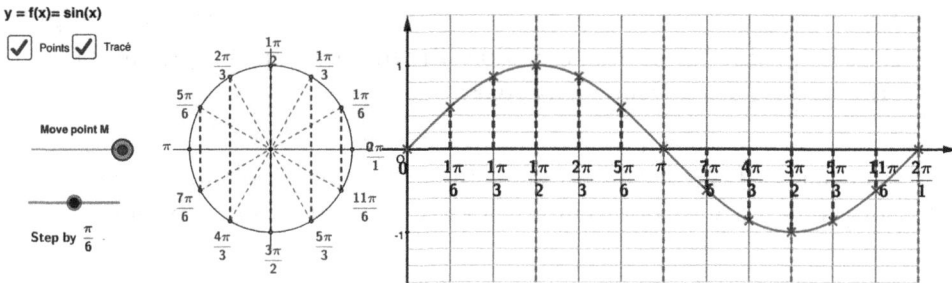

Figure 2.16 Connected dynamic geometry and dynamic graphing calculator representations.

tools and decide that the variety of representations using both continuous and categorical data available in a tool like CODAP would be helpful and could then import the data easily into this second technology for continued exploration. Regardless of whether you choose one technology or more to use, the goal should always be to select the technology that includes the types of mathematical objects that you want to position students to act on and reason about. In the coming chapters, we will learn more about how teachers who use these technologies regularly make such decisions.

CHAPTER TAKEAWAYS

- Math action technologies are different from other general educational technologies – they allow students to actually act on math objects and observe how those objects react (in mathematically correct ways) to their actions.
- There are many different types of math action technologies, as a secondary teacher it is important to get comfortable with technologies that allow for exploring algebra, function, geometry, and statistics/probability.
- Activity builders can provide a structure for your lessons and provide you with a means to monitor student work and select student work to make public to foster mathematical discussion.

Questions to Discuss With Your Colleagues

1. Looking across the examples provided for how each technology type might be used in a secondary math class, what do you notice?

2. For each of the technology types described in this chapter, how might it be helpful for students to explore the mathematical topics in the courses you are teaching this year?

As a reminder, you can see a list of specific technology tools in each category in the supplementary links provided in Link 2.1 earlier in the chapter.

 You can find links to all the technology-enhanced tasks and supplementary videos throughout the book at https://www.tlmtresearch.com/teachingmathtechbook.

3

Deciding When and What to Use

One of the most common questions we are asked is, "How do I know when to incorporate technology in a lesson?" When planning for instruction, whether or not to incorporate technology in a lesson, and if so which technology, are among the first decisions to be made – after determining your learning goals, of course. This decision is not trivial. It is easy to get pulled into using technology because on the surface something looks really cool or fun or because you have to show you use technology on an evaluation rubric. When thinking about how he decides when to use technology, Andrew recalled his first years of teaching:

> When I first started teaching technology was a nice new cool thing and I was like, oh this is just how we're gonna engage people. Right? We're just gonna throw in technology and it's just gonna work. But deciding when to use it and when not to use it is almost as important as the technology itself.

We agree with Andrew, that is an important decision. The purpose of this chapter is to provide some guidelines to help you decide whether to incorporate technology into a specific lesson. Along with guidelines for when you do decide to use technology, we also discuss determining which type of technology might be the best choice.

DOI: 10.4324/9781003302285-4

Deciding Whether or Not to Use Technology in a Lesson

Like any other lesson planning decision, deciding whether to use technology in a lesson should be driven by your learning goals. With your learning goals identified it is often suggested that you ask yourself if technology will make the lesson better. But what does that mean? When we asked our Tech-Math Teachers about this, they said things like,

- "I think I always just start off with wondering what my goal is as an entry point and trying to determine what technology would be best served in what way – because I know there is a really big push for teachers to use technology, but I think that we need to be intentional about what we choose to use and when". (Kristy)
- "I'm trying to think about, 'Hey, is this saving time?' 'Hey, has this helped me portray a key idea to my students?' It's one thing for us to like technology as teachers, but we want to make sure that when we put it in our students' hands they're using it for increasing their learning and creating a valuable learning experience". (Nick)

Keeping in mind that our goal is to use technology to position students as math explorers. That means considering the ways technology might remove barriers and provide students agency.

Math action technologies afford the ability to create and explore representations, meaning that technology can create a valuable learning experience when it is used for considering multiple representations before generalizing, seeking out diverse or extreme examples, and attending to definitions (Belnap & Parrot, 2020). With this in mind, we recommend asking yourself the following questions when deciding whether to use technology in a particular lesson:

- Will technology allow us to see the mathematical concept of interest differently?
- Will technology allow us to create representations quickly and accurately so we can focus on analyzing and/or generalizing?
- Will technology open us up to different solution strategies?
- Will technology provide meaningful and immediate feedback?

In the following sections, we dig into each of these guiding questions, hearing from the Tech-Math Teachers and examining example tasks that help to illustrate how they can guide your decision about whether to use technology in a particular lesson.

Will Technology Allow Us to See the Mathematical Concept Differently?

Given the abstract nature of mathematical ideas, representations are how we gain access to the ideas. Representations provide us something to point at, critique, and discuss. Technology can be helpful when it allows us to create representations that are connected and dynamic. As we noted in Chapter 2, what sets math action technologies apart from other technologies is that they allow us to create and act on mathematical objects, often in dynamic ways. In fact, many technology tools allow us to create multiple connected representations (e.g., a table, equation, and graph that are all connected – changing one results in a change in each of the others), something that is not possible with paper and pencil. It is this connection between idea and representation that teachers say they think about when considering if technology will allow them to see a mathematical idea differently.

- "I'm going to use technology when I want them to move between mathematical representations". (Nathan)
- "If we're only working with symbols and I feel like I'm gonna have students who struggle, then that gives me another medium where they don't just have to attach to a symbol, they could look at a table or a graph or a picture". (Sarah)
- "I tend to use it when something has a lot of complex *mathiness* that we can simplify with a picture or an animation where they can get the gist of it before we actually dive into the math". (Andrew)

Ultimately, if students' insight into the mathematical idea would be enhanced by the creation and analysis of multiple connected representations, then it would make sense to incorporate technology in the lesson.

An Example: Pig Pen

Here is a traditional problem found in many curricular materials that can be used from middle school up through calculus.

> A farmer wishes to construct a rectangular pen for her pigs. She has 48 meters of fence and wants to build a pen that has the largest area possible. What would be the dimension of this pen?

PAUSE AND CONSIDER

1. What representations would you expect students to create when they are working on this task?
2. What insight into the situation does each representation provide?
3. What connections do you hope students can make among the representations?
4. How might dynamic connected representations support students' sensemaking?

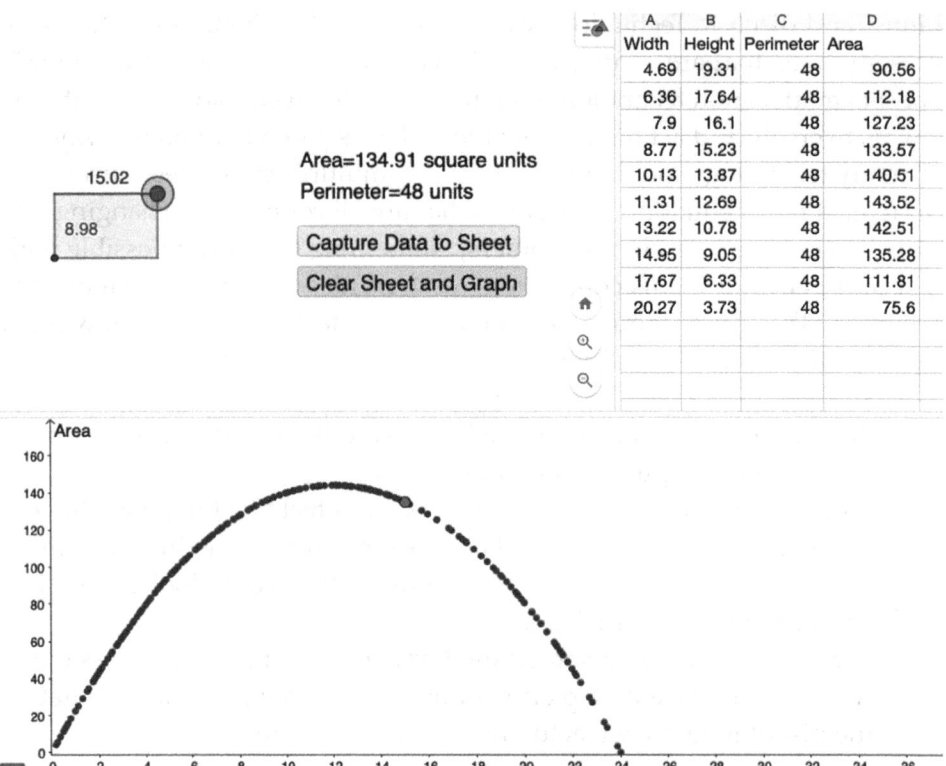

	A	B	C	D
	Width	Height	Perimeter	Area
	4.69	19.31	48	90.56
	6.36	17.64	48	112.18
	7.9	16.1	48	127.23
	8.77	15.23	48	133.57
	10.13	13.87	48	140.51
	11.31	12.69	48	143.52
	13.22	10.78	48	142.51
	14.95	9.05	48	135.28
	17.67	6.33	48	111.81
	20.27	3.73	48	75.6

Area=134.91 square units
Perimeter=48 units

Capture Data to Sheet

Clear Sheet and Graph

Figure 3.1 GeoGebra building a pig pen applet.

Rather than jumping right to a symbolic representation of the situation, providing students with a tool to explore using multiple connected representations can provide them with additional insights. For example, using GeoGebra (Figure 3.1 & Link 3.1) we can construct a file in which students can drag the corner of the pen diagram and see that the perimeter remains 48 units while the length, width, and area are all changing with respect to each other. The table of values – which was created by capturing values from the image as the corner of the pen was dragged – provides values with which to look for patterns and the graph suggests that at a certain point, the area becomes the largest possible and then starts to decrease again.

Link 3.1 Building a Pen GeoGebra Applet

There is also a similar task created by Desmos that uses Desmos Activity Builder (Figure 3.2). Again, students change the size of the pen (or in this case a field with a perimeter of 50 m), capture data from the figure into a table, and examine the connected graph. The representations are dynamically connected in both examples – side by side in GeoGebra, on separate pages in Desmos. Both are excellent interpretations of the task that take advantage of what technology

Build a Field

Now try to make the biggest field you possibly can that still only uses 50 meters of fencing.

Press the button one more time when you are done.

Build Field C

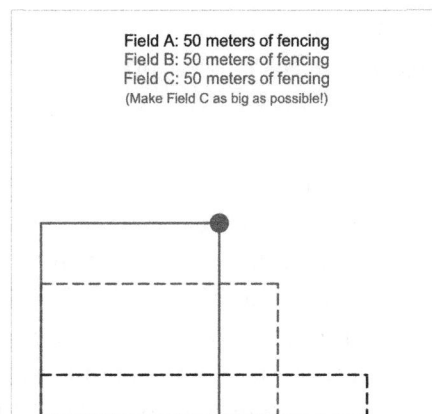

Field A: 50 meters of fencing
Field B: 50 meters of fencing
Field C: 50 meters of fencing
(Make Field C as big as possible!)

Calculate the Area of Your Fields

These are the three fields you created.

You said that Field C will have the greatest area.

Calculate the width and area of each field to see if you were right.

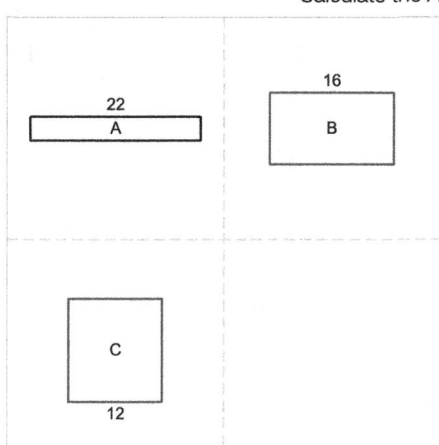

Field	Length (m)	Width (m)	Perimeter (m)	area (sq. m)
A	22	3	50	66
B	16	8	50	128
C	12	13	50	156

Graph the Data

This graph shows all the fields our class created today.

Your points are in orange. Your classmates' are in blue.

What function family do you think best describes the data?

○ Linear

○ Quadratic

○ Exponential

○ Other

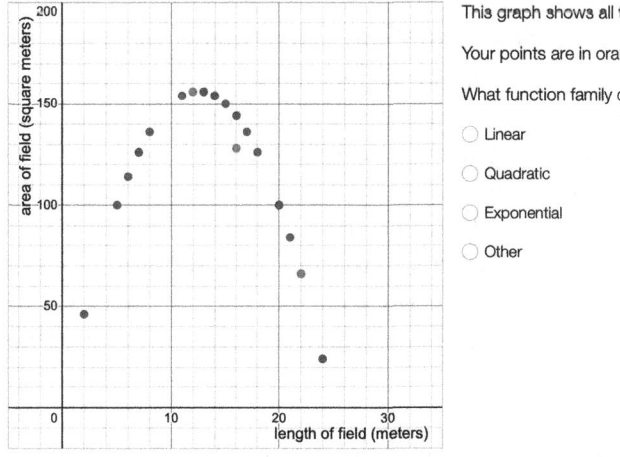

Figure 3.2 Connected representations in the Desmos building a field activity.

can offer that supports student thinking about the relationship between the context and the various representations. (Note: See DePeau & Kalder, 2010, for a similar task and related student work using the TI-Nspire platform.)

Link 3.2 Building a Bigger Field Desmos Activity

The dynamic aspect of the connected representations provides insight that is difficult to recreate with pencil and paper. As students drag the corner of the pen, they can see the difference between the values for the length and width decreasing in the table as the area of the pen increases (and the difference increases as the area decreases). In addition, dragging the corner of the pen at a constant speed and watching the graph one would see the point being traced decrease in speed as it gets close to the value of the maximum area and then begin to increase again as it moves farther away (i.e., they can observe the increasing and decreasing rate of change). If students conjecture that the maximum area occurs when the value of length and width are equal, they can easily test their conjecture by changing the pen and observing the results, by scrolling through the table of values, or by closely examining the graph. In fact, the ability to zoom in on a graph is a powerful tool.

- "I love that we can sit there and we can just keep zooming in and zooming in and really see that behavior of the graph of approaching and getting infinitely close and really talk about it and trace the x and y values". (Nolan)

Here zooming in at the maximum of the graph and clicking on the maximum will result in showing the ordered pair of the point. One can imagine how using the zoom feature could be very useful when exploring graphs of all types.

Will Technology Allow Us to Create Representations Quickly and Accurately So We Can Spend Time Analyzing Them and Working to Generalize?

While it is sometimes helpful to work with multiple connected dynamic representations, other times you might want students to be able to quickly and accurately create one or two representations (this is sometimes referred to as using the technology as an amplifier; Pea, 1985). Whether it is a geometric construction or the graph of a function, creating representations quickly and accurately allows more time to focus on sensemaking – a use of technology that many of the Tech-Math Teachers indicated as being very valuable to them and their students.

- "I learned about how to graph parabolas by graphing a zillion parabolas on graph paper. … I just remember graphing a lot. So, when we learned about translations of parabolas, we would make a table for $y = x^2$, and then we make a table of $y = (x - 2)^2$. We do a zillion tables and zillion graphs. That can be so much better with technology. Technology is removing a lot of the headaches and it's really providing us a way to quickly access the concept at hand. The concept at hand is not filling out a table. The concept was function transformations. So, when I see that kind of topic, when I have a topic where technology is going to reduce what the student needs to do so they can just access the math more directly, that feels like a really good time to use technology". (Dan)

- "In geometry it would be something that allows students to discover the properties of a shape through movement or measurement. Things that they can experiment with dynamically whereas on a piece of paper they'd be measuring and cutting. I did that for years and they make mistakes, they make a mess, they measure wrong, they get the wrong data so it makes it easier to get the right data and come up with the conclusion faster". (Michele)

Many teachers shared stories similar to Dan's and Michele's, emphasizing that they choose to use technology when it can remove the headache of creating a lot of examples of similar representations so the focus can be on the mathematical idea, not the process of creating the representation. In sum, when creating a lot of examples quickly and accurately is your goal, then technology can help you accomplish that.

At the heart of deciding whether to use technology to create representations quickly and accurately is whether you intend to position your students as explorers that use those quickly created representations. In fact, sometimes taking the time to create representations by hand can pull focus from the analysis and generalization process.

- "If I want them to be exploring a concept and discovering concepts, then I'm probably going to use some form of technology to really emphasize that idea. Let's go discover patterns and relationships. That's the way I frame my classroom from the very beginning. Mathematics is the study of patterns and relationships, so we need to be able to notice them. We need to be able to describe them and generalize them. So, when that's the primary focus, let's use technology to help speed up that process". (Nathan)
- "Like if they're looking at a family of functions and they have to graph all of them by hand to see the transformation and how it moves, they kind of get lost in the process and don't really get to analyze like they should. It's all about the analysis of the thing, not getting to the thing to analyze". (Andy)

Suppose ABC is a triangle. Let D be the midpoint of $\overline{AB}$ and E be the midpoint of $\overline{BC}$.

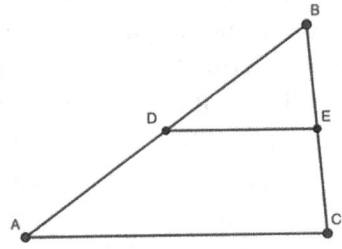

a) Show that $\overline{DE}$ is parallel to $\overline{AC}$.

b) Show that the length of DE is half the length of AC.

Figure 3.3 Triangle midsegments task.

When talking about creating representations quickly and accurately many of the Tech-Math Teachers used graphs as examples, but this principle applies to geometric constructions as well. Constructing figures using dynamic geometry technology has the same effect since dragging the figure results in creating many examples to compare and contrast.

An Example: Triangle Midsegments

In Figure 3.3 is a traditional geometry problem found in many curricula materials. The intent is to introduce students to a theorem about the relationship between the length of a triangle midsegment and the side to which it is parallel.

PAUSE AND CONSIDER

1. How might using technology to create numerous example triangles (or one that is dynamic and can be dragged) change the nature of this task?
2. What insight into the context would a dynamic construction of the image provide?
3. What questions might you pose with this task if you were to use technology?

Students can quickly create a triangle from three line segments, use the midpoint tools to add a midsegment, and then measure the midsegment and the side. Then by dragging the vertices of the triangle, they can quickly create a multitude of examples to explore and look for patterns (Figure 3.4). They will likely quickly notice that the length of the midsegment is half of the length of the side. Does that appear to be always true? If so, why is that

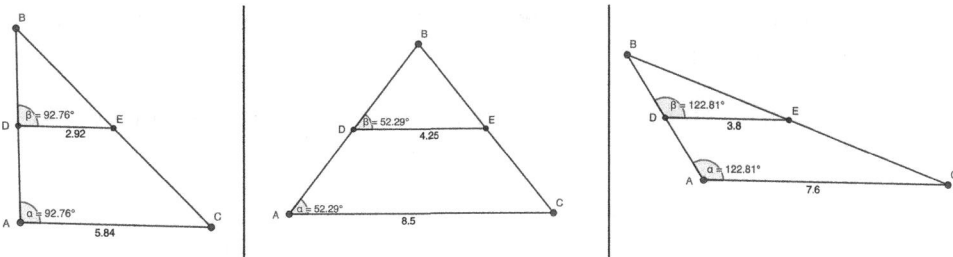

Figure 3.4 Constructed triangle midsegments.

always true? Exploring the dynamic construction can provide key ideas to the proof. For example, they will likely notice that the midsegment appears to always be parallel to the triangle side. They can test this conjecture by either measuring the slope of each segment or by measuring corresponding angles. Raman and Weber (2006) note that the intuitions that students build through explorations like this can lead to identifying key ideas to proof. In this case, the key idea is that the midsegment is parallel to the side. Thus, we have similar triangles in which the side lengths are proportional by a factor of two.

Technology makes it relatively easy to transform traditional curriculum materials into opportunities to explore. When representations (including geometric constructions) are easy to create and interact with, students will naturally notice and wonder. Many of the Tech-Math Teachers shared that they decide to include technology in their lessons when they want to take an inquiry approach to a mathematical idea.

- "If I'm trying to get them to pick up on a pattern of something that's happening over and over, some kind of mathematical relationship, technology generally allows them to experiment and conjecture and confirm those ideas a lot faster and with a lot more confidence. It gives kids the opportunity to find out this stuff for themselves, rather than me just saying hey this thing exists. They can kind of uncover it on their own, which gives them a little bit more sense of agency over the mathematics that they're developing". (Kristen)

Will Technology Open Us Up to Different Solution Strategies?

Being able to create representations quickly and accurately can also open students up to different solution strategies. Emphasizing multiple solution strategies and representations and making connections among them not only position students as competent math explorers (Bonner, 2021; Suh et al., 2022) but also support students' learning (NCTM, 2014). Many of the Tech-Math Teachers noted that sometimes including technology in a lesson is not done in a structured way; rather, they decide to have it available to any students that would like to use it.

- "I can't really think of a time when I would completely remove technology from their hands". (Nathan)

By having technology available for students to use as they see fit, they are free to create representations that might introduce different solution strategies for the class to discuss and learn from. By being open to different solution strategies, we are being open to voices that might otherwise not have been heard in our mathematical discussion (Suh et al., 2022). Student voice and student agency are important considerations when thinking about whether to include technology in a lesson. The Tech-Math Teachers not only noted the importance of providing space for student voice but also the joy of students using strategies they didn't anticipate and getting to share their brilliant ideas with the class.

- "I think about student voice. Is the technology going to increase student voice?" (Zach)
- "And then they become the mathematician, right? They have shown everybody else. This is really how this will work, and I've come up with this by myself. So, it's a fantastic way of validating them as yes, I'm a mathematician, too. I'm making sense out of this problem". (Karen)

An Example: The Zip Line Task

A new amusement park is building a zip line attraction. The attraction will have two towers on opposite sides of a man-made lagoon full of alligators. The lagoon will be 600 m wide. One tower will be 100 m tall, and the other will be 60 m tall. There will be two zip lines, one from each tower, that riders will take from the tops of the towers to an island in the lagoon. Once on the island, riders will exit the ride by walking across a long bridge. But zip line wire is expensive! How far from the bank of the lagoon should the island be in order to minimize the length of the zip line wire? (Wilson et al., n.d.)

PAUSE AND CONSIDER

1. What strategies (both tech and non-tech) do you expect students might use to determine the minimum length of zip line wire needed?
2. Specifically consider the ways that students might use spreadsheets, computer algebra systems/dynamic algebra notation systems, dynamic graphing technology, and dynamic geometry technology to make sense of the zip line task.

The zip line task is accessible to students who are familiar with the Pythagorean theorem through calculus, and it can be solved many different ways. Among students in a second-year Integrated Math course, common solution strategies include using the Pythagorean theorem to create tables, equations, and graphs or using similar triangles to set up ratios and proportions. When technology is available students use similar strategies but often more efficiently and accurately. In addition, we also see other creative strategies come out (Figure 3.5). For example, one student set up a spreadsheet with formulas for each of the wire lengths and then filled the columns to find where the total was minimized. Notice that after identifying 375 m as the minimum, they added rows to the bottom to check values closer to 375 m as well. A common strategy is to set up an equation and then use the graphing calculator to determine the minimum. Another student used dynamic geometry technology to build a simulation. Dragging the point between the towers, they were able to approximate where the amount of wire would be minimal. Another did something similar but, to make sense of the situation, reflected one of the triangles created by the tower, wire, and ground/water and determined that the wire would be minimal when the two hypotenuses "fell in a line". Brilliant ideas! Together these strategies along with paper-and-pencil strategies provide opportunities for great mathematical discussions.

Will Technology Provide Meaningful and Immediate Feedback?

One final consideration when deciding whether to include technology in a lesson is whether it can provide students immediate and meaningful feedback. When considering math action technologies, we aren't talking about entering an answer and getting a green checkmark letting you know it is correct (although that is valuable at times too), but rather, about students creating a representation that provides them feedback about their ideas or solutions. Zach shared a great example of this. His students were struggling with the difference between intercepts and intersections. Knowing that they were struggling with the difference he used something similar to the following task.

An Example: Intercepts vs. Intersections

Find the x- and y-intercepts for each of the following functions; then find their points of intersection. $f(x) = x^2 - 2$ and $g(x) = x + 4$.

Zach knew his students could determine the intercepts and intersections using paper-and-pencil procedures, but they often got confused on which was which and sometimes wrote the ordered pairs backward. By simply making technology available to those that would like to use it, his students could test their results by graphing (Figure 3.6). He explained:

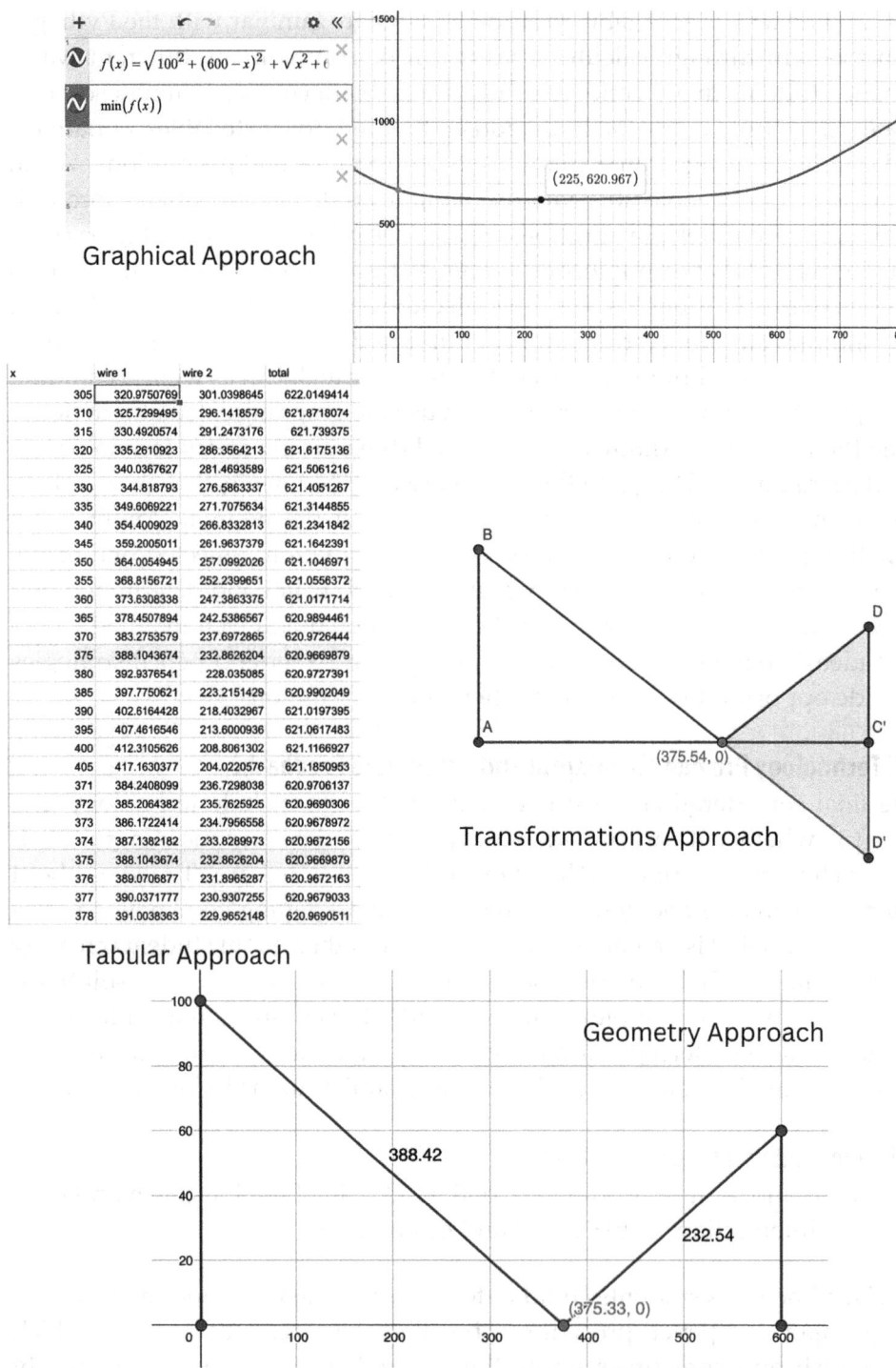

Figure 3.5 Multiple strategies for solving the zipline task using technology.

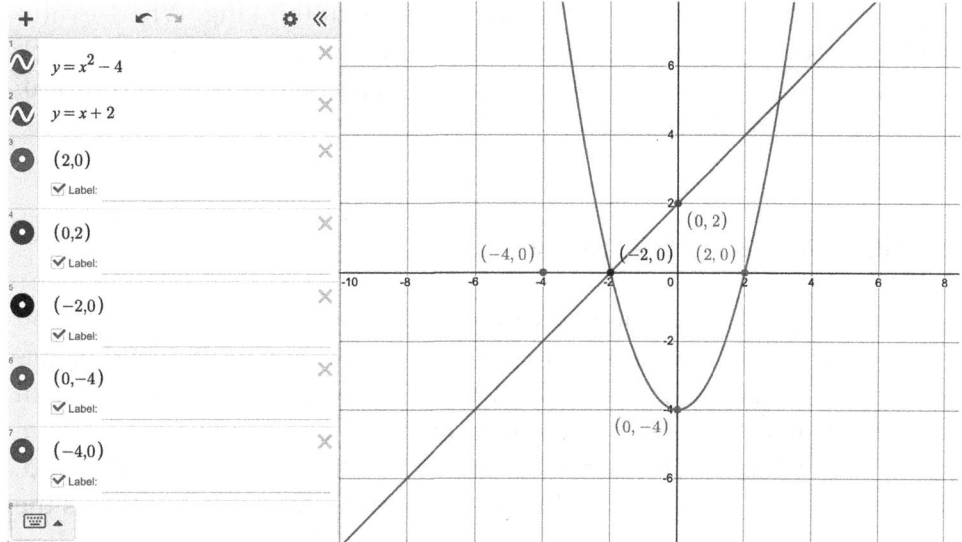

Figure 3.6 Graphing to check intercepts vs. intersections.

Students can self-check by throwing up the graphs, and type those points in, and they show up on the graph. right? So, if they're being asked to list the intercepts of the line they can type in (–4,0) and (0,4), and see that there's actually a matchup. The mistakes that you would get on paper is they're typing in (y, x) instead of (x, y). They can easily see when they type that in. 'Oh, no, that's not right. Let me fix my x and y.' Or they'll write down points of intersection without realizing it. They type those in, and those points pop up and show. Oh, those are actually where they intersect. Right! I want the intercepts.

In this situation, although students might have been wrong, they got immediate feedback – not from the teacher but from their own actions – there was no judgment but an opportunity to refine their thinking and revise their answers.

In Zach's example, students could have used any graphing technology to check their work and receive meaningful and helpful feedback. In fact, graphing technologies are commonly promoted as tools for checking work done by hand, and students greatly value being able to use them in this way (e.g., Doerr & Zangor, 2000; McCulloch et al., 2012). Students can use statistics and geometry tools to check their work as well. What is important is that they have access to appropriate tools to use in ways that provide them meaningful feedback. In fact, feedback is so valuable that Desmos has identified it as one of the important principles that guide the design of their technology-enhanced activities. They state, "Whenever possible, we use computers to interpret rather than evaluate student thinking. Where evaluative feedback says, 'you're right' or 'you're wrong', interpretive feedback says, 'here's how we're understanding your answer. What do you think now?'" (Moynihan et al., 2021).

Deciding Which Technology to Use

If you've decided that technology would make your lesson better, now you need to decide which technology to use. Once again, we recommend beginning with your learning goals, as this will help you to quickly narrow down your options. In Chapter 2, we described the types of technologies you have to choose from. Most of the technology types are aligned with a math strand, so knowing your goals narrows down your options dramatically. This is the first pass that many of the Tech-Math Teachers said they took when deciding which technology to use.

- "A lot of it has to do with the strand. So, like if it's stats I'm probably going to go Desmos or CODAP because those are going to be more specific to the strand in which I'm trying to apply it. Same thing with geometry. I'm probably going to go to GeoGebra because the affordances there make more sense with what I'm trying to get kids to understand". (Kristen)

Once you have narrowed down your list of options, it is helpful to start with some general questions: What technology do I have access to? Do my students need an account? Is it user-friendly? We think of these questions as assessing barriers. For example, if a technology isn't free and my school doesn't have an account, then there is a barrier to my using it.

- "I think the first thing – is it free? Do I need a subscription? Is it even available? What's the accessibility for me as the teacher but also for my students?" (Nolan)

If students need accounts, you'll need to think about the time it takes to set that up. This is often very worthwhile if it is a technology that you will use a lot throughout the year, but if it is for one-time use, then maybe not. Finally, if you are going to spend more time teaching how to use the technology than having students explore mathematics, that is a barrier to consider as well. Again, if it is a technology that you'll use a lot throughout the year, it is likely worthwhile.

Driving the decision about whether to use technology at all was the role that representations might play in the lesson. With that in mind, it is important that when you select a technology to use that it can be used to create the representations needed in the lesson. For example, to create an ellipse using the Desmos graphing calculator, we are restricted to using the equation to create a graph. In GeoGebra, since the geometry tools are available in the graphing view, we can either graph an ellipse using the formula or construct it using its geometric definition. In addition, we need to check and make sure that the representations are mathematically correct. While most math action technologies available today are mathematically correct, this step is especially important when considering preconstructed interactive applets.

When it comes to symbolic, tabular, and graphical representations the symbolic is the biggest difference among the various dynamic graphing technology options. For example, when graphing a linear equation in some technologies we can simply enter $y = mx + b$ and the technology will automatically create sliders for m and b and graph the resulting line. In others, you can only graph specific linear equations, like $y = 2x + 1$. Still others must assume an equation is a function to graph and either require it be entered in function notation (e.g., $f(x) = 2x + 1$) or will automatically assign a function name to the equation that is entered ($f: y = 2x + 1$). If students are familiar with function notation, any of these is an appropriate choice. If they are unfamiliar, then it is possibly best to select one that does not require or assign it so as to avoid confusion. However, if one of your goals is to support students developing an understanding of function notation, then you might want to select a technology that specifically does use or assign function notation.

For statistics lessons, your choice of representations is dependent on not only the representations you plan to use but also the nature of the data set and how you want students to be able to interact with the data. If students will explore a data set that includes many variables, and variables of different types (e.g., categorical, quantitative) then a technology like CODAP might be helpful as the entire data set can be pulled into CODAP at one time and students can create a plethora of representations to look at relationships among variables and create statistics to support their exploration. However, if the goal of your lesson is to use a list of numerical data and accompanying representations to help students build an understanding of particular statistical measures (e.g., formulas for measures of center or measures of spread), then a dynamic graphing calculator application may be more appropriate as it will allow students to act on individual data observations and observe the changes to the statistics as a result of their actions.

> **PAUSE AND CONSIDER**
>
> Imagine you are getting ready to plan a lesson focused on coordinate quadrilaterals (i.e., given four points describe the quadrilateral and determine its area and perimeter) and have decided that technology would be good to include as it would allow students to explore their conjectures and solutions using both geometric and algebraic constructions.
>
> 1. What tools would you want to make sure the technology you select includes?
> 2. Open at least two different dynamic geometry tools and consider which would be the better selection given your goal.
> 3. Is there anything else you would take into account when making this decision? Explain.

If you have carefully considered the representations, you anticipate being important to your lesson – both what they look like and how you and your students would create them – and find more than one technology fits the bill, then we recommend selecting the one that you and your students are most comfortable with. For those of you that have a lot of experience with particular technologies, you might follow the advice of some of the Tech-Math Teachers and start where you are most comfortable.

- "Personally, I always start at Desmos and then if Desmos can't do what I want it to then I find something else". (Samantha)
- "I have a classroom set of calculators [TI-Nspire], so if I can do it with them, then I'll do it with them. But if I want the kids to manipulate a shape it's just harder to do on there". (Andy)
- "So, my first layer is going to be is it a tool that I already know how to use. If it's not a tool that I already know how to use, but I really think it's awesome and worth using I might put the time in to do it". (Lauren)

Starting with your favorite technology, see if the representations you wish to create are possible and if they are appropriate for your students. If they are, great, then decision made. If not, then consider other options appropriate for the math strands your lesson is addressing.

Using Math Action Technologies within Activity Builders

Once you have decided which technology best suits your needs for a lesson, an additional consideration is whether or not to embed it in an activity builder. As we explained in Chapter 2, many technologies have platforms for developing activities that can include math action technologies alongside ways for students to communicate their ideas (i.e., text or drawing) and for teachers to monitor, provide feedback, and display student work. We do not recommend automatically using an activity builder; rather, think carefully about what it adds. On one hand, activity builders can help with pacing through a lesson and provide an opportunity for you to see all students' ideas and to share them publicly. On the other hand, it can take a lot of time to build a high-quality activity, and when not well designed, they can sometimes restrict student creativity. The Tech-Math Teachers articulated how they weighed these decisions, thinking about when a clearly articulated task on a sheet of paper accompanying the technology outweighs using an activity builder and vice versa.

- "Really, if I want to stop them at certain points and have discussion, that's going to lend itself towards Desmos Activity Builder, because I can pace it and pause them and stop and then give instructions on the next step. In the Desmos graphing calculator I can have them organize their work in folders. I can have them build tables, build functions, and so I can have a lot of different things going on there. And I like the creative aspect of it. Who's authoring this work? Who's got the authority in this piece here? That is what I think about when deciding between the two". (Sarah)
- "So, an example. I have some coordinate geometry challenges. Like create a polygon with these coordinates. Dilate the polygon by a factor of, you know. Translate the polygon down four. And so that's a document with just a list of challenges and they use their graphing calculator app to go through those challenges. It's a little bit easier for the students to just kind of go through that way instead of working through activity builder". (Leah)

Hopefully you now have a clearer idea about how to decide when to use technology in a specific lesson. In Chapter 4, we discuss specific guidelines for selecting, adapting, and/or designing tasks that align with your learning goals. Here are a few parting thoughts from the Tech-Math Teachers:

- "You've got to be careful with technology, because sometimes you get wowed by the technology and you're like oh this is the coolest thing. But you should really pick technology that helps you get more insight into your students' thinking or lets the students become more of thinkers. Don't pick technology that makes their lives easy or that they just have fun playing with. There's a time and place for that, but I'm using technology as an instructional tool". (Nick)
- "How do I use the technology to enhance their thinking? So, the biggest thing that I ask myself is if they're understanding of what's going on is just surface. Then we've got to go deeper. The deeper typically prompts you to pull out the technology". (Sarah)

CHAPTER TAKEAWAYS

When thinking about whether to use technology in a specific lesson, ask yourself these questions:

- Will technology allow us to see the mathematical concept of interest differently?
- Will technology allow us to create representations quickly and accurately so we can focus on analyzing and/or generalizing?
- Will technology open us up to different solution strategies?
- Will technology provide meaningful and immediate feedback?

If you decide technology is appropriate for your lesson, as you work to select a specific technology ask yourself these questions:

- Can I create the representations we need?
- Is the form of the representation appropriate for my students?
- Is the mathematics accurately represented?
- Is the technology easy for us to use?
- Do I want to embed the math action technology into an activity builder?

Questions to Discuss With Your Colleagues

1. What types of learning goals do you think lend themselves to technology use?
2. What keeps you from using technology even when you know it would support student learning and mathematical identity development? What would make you feel more confident in using it?

3. What are your go-to technologies? What representations do they do really well? Which do they do less well? Are there other, more appropriate choices when representations that the technology does less well are important to support student learning?
4. What is a task you have used recently that you now think might be a good one to revise so that it incorporates technology? Explain.

 You can find links to all the technology-enhanced tasks and supplementary videos throughout the book at https://www.tlmtresearch.com/teachingmathtechbook.

4

Selecting Technology-Enhanced Tasks

The tasks you select to use in your lessons determine the nature of the math your students will engage in. If you have decided that technology is a good choice given your learning goals, then the next step is to carefully select the specific task or sequence of tasks to use in your lesson. These choices greatly impact your students' learning, the development of their mathematical habits of mind, and their identity as math learners.

First and foremost, when selecting a task, it needs to be aligned with your learning goals. (Yes, we've mentioned this before, but it is worth repeating.) It is easy to get caught up in the excitement of finding a "cool" technology task on the topic of your lesson, but if you aren't careful, you might end up selecting a task that, while on the right topic, does not support your intended goals. For example, the Desmos activities What Comes Next (Link 4.1) and Predicting Movie Ticket Prices (Link 4.2) are both focused on exponential functions (and comparing them to linear functions), but they are designed for different learning goals. What Comes Next is intended to support students' learning about the ways that exponential functions grow differently than linear functions through investigating tables and graphs and how to determine an exponential equation given its graph (Figure 4.1). In contrast, Predicting Movie Tickets is focused on applying what students know about linear and exponential functions to build a model for the cost of movie tickets in the US over the last 100 years (Figure 4.2). The What Comes Next and Predicting Movie Tickets activities have a lot in common with the dynamic representations they use and the concepts they address, but they have very

DOI: 10.4324/9781003302285-5

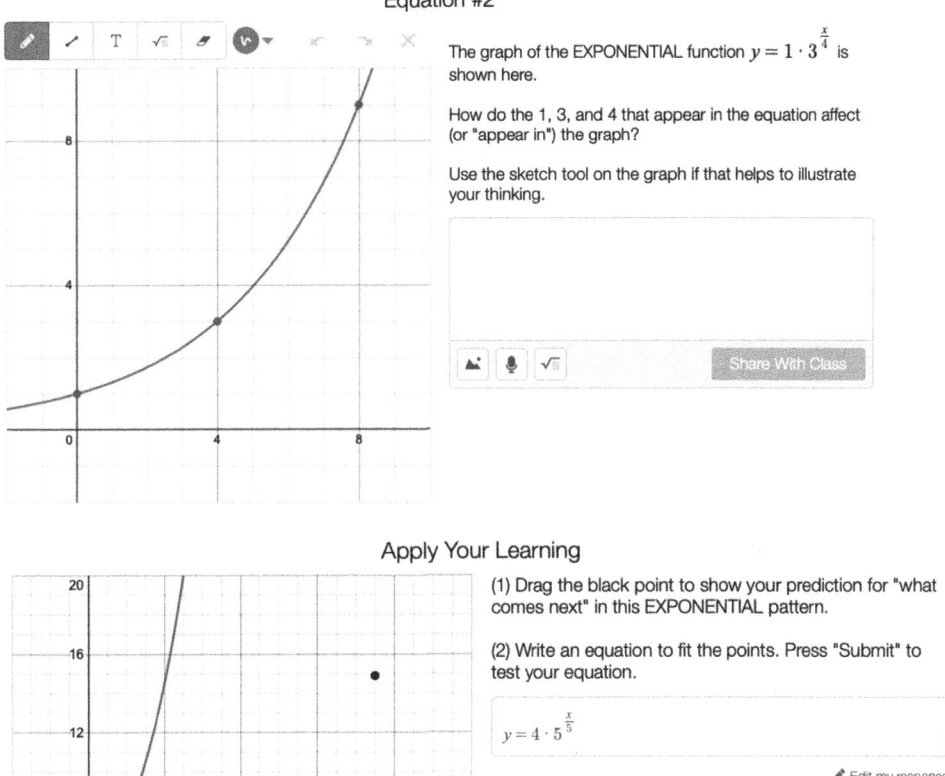

Equation #2

The graph of the EXPONENTIAL function $y = 1 \cdot 3^{\frac{x}{4}}$ is shown here.

How do the 1, 3, and 4 that appear in the equation affect (or "appear in") the graph?

Use the sketch tool on the graph if that helps to illustrate your thinking.

Share With Class

Apply Your Learning

(1) Drag the black point to show your prediction for "what comes next" in this EXPONENTIAL pattern.

(2) Write an equation to fit the points. Press "Submit" to test your equation.

$$y = 4 \cdot 5^{\frac{x}{5}}$$

✏ Edit my response

Figure 4.1 Desmos activity – what comes next.

different learning goals. This highlights the importance of weighing the great tasks you find against your learning goals to make sure they align well.

Link 4.1 What Comes Next

Link 4.2 Predicting Movie Ticket Prices

Even with your learning goals guiding the way, selecting a technology-enhanced task can be a daunting mission. Our goal in this chapter is to provide you with some guidelines for selecting a task – and, when you can't find the perfect one, for adapting or building one of your own.

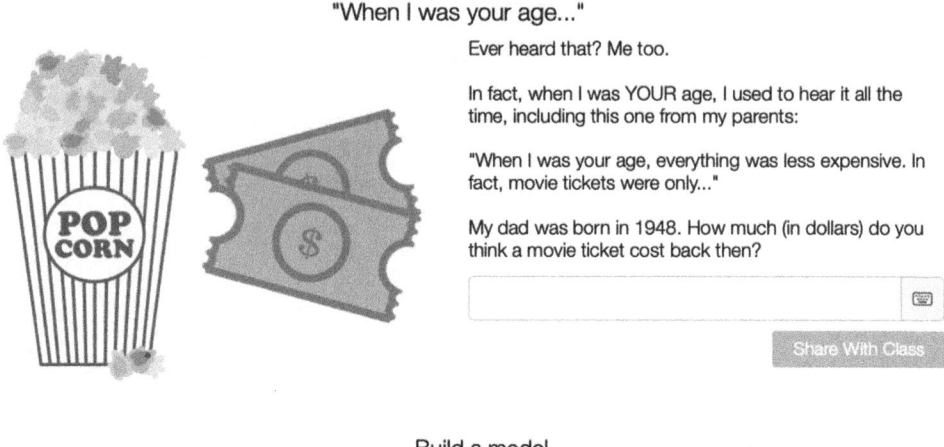

"When I was your age..."

Ever heard that? Me too.

In fact, when I was YOUR age, I used to hear it all the time, including this one from my parents:

"When I was your age, everything was less expensive. In fact, movie tickets were only..."

My dad was born in 1948. How much (in dollars) do you think a movie ticket cost back then?

Share With Class

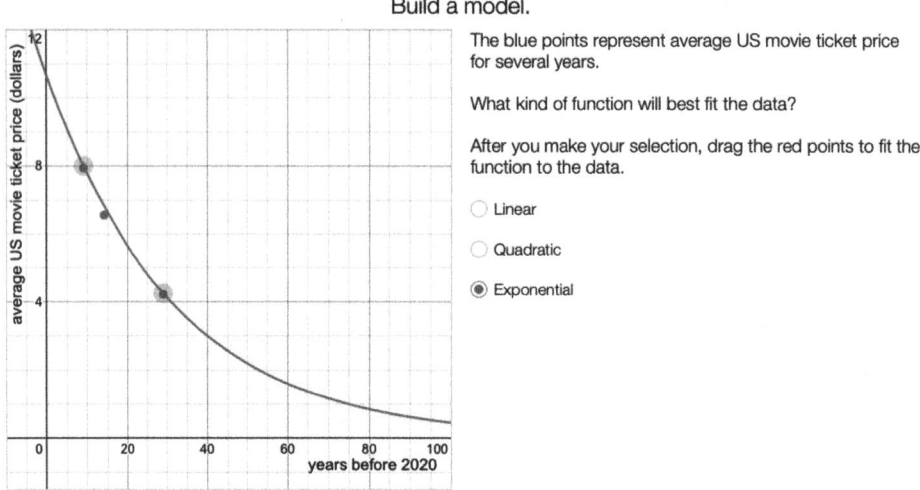

Build a model.

The blue points represent average US movie ticket price for several years.

What kind of function will best fit the data?

After you make your selection, drag the red points to fit the function to the data.

○ Linear

○ Quadratic

◉ Exponential

Figure 4.2 Desmos activity – predicting movie ticket prices.

Characteristics of a Good Technology-Enhanced Math Task

Technology-enhanced math tasks have two equally important components – the math action technology itself and the accompanying prompts. As we noted in the introduction to this book, we think of these two components collectively as task + technology. The technology is sometimes an integral part of the task, and other times, it sits alongside the task. Either way, the technology determines the representations that students can engage with, and the accompanying prompts will drive how students interact with the technology – the potential nature of their thinking.

As we noted in Chapter 1, a good technology-enhanced math task is often described as being "low floor–high ceiling–wide walls". The notion of a low floor (anyone can access the task) and a high ceiling (there are lots of possibilities for taking things even further; Papert, 1993) has been a principle for designing experiences with learners and computers since the 1970s. The notion of wide walls was added to express the importance of learners being able to "explore multiple pathways from floor to ceiling" (Resnick, 2016). When asked about the characteristics of a good technology-enhanced math task, the Tech-Math Teachers used this language to describe what they are looking for.

- "I'm looking for low floor, for high ceiling". (Zachary)
- "A low floor entry, but we want that high ceiling there. So it's like so anybody can start the activity". (Dan)
- "When they jump into it, I don't want there to be any barriers, you've got to be able to get right into it. So, it should be [a] very, very low floor. And then the benefit of the technology is that you have all these representations at your fingertips … and there'[re] going to be multiple entry points, too". (Sarah)

An Example: Graphing Stories

A common curriculum task asks students to either interpret a graph given a context (e.g., a bike ride, a walk to school, a container being filled with water) that is changing over time or draw a graph that matches a context. The purpose of these tasks is to emphasize dependent and independent variables, and covariation and create graphs that model real situations. For example, consider this task:

> Claire is at the circus, and one of the acts is a person being shot out of a cannon, straight up into the air. She was scared at first, but luckily, they are wearing a parachute. The parachute opens as the person begins to fall back down to the ground. They immediately slow down and fall gently back down to the earth, landing safely on the ground.

Using paper-and-pencil techniques, students would be asked to sketch a graph that represents the person's distance from the ground over time. But does that allow them to really explore how the variables (distance and time) actually change together? Does it provide feedback to them on their ideas about how a graph might relate to the situation?

Now consider the Function Carnival activity from Desmos (Link 4.3). In this activity, students explore three different animations – a person being shot out of a cannon with a parachute, a bumper car driving among other bumper cars, and a cart going around a Ferris wheel – each situation more complex than the prior. In each situation, students can sketch a graph of the identified distance vs. time and then play the simulation to see how their graph is represented in the changing distance as time passes and compare how their graph would result in the animation (Figure 4.3). The activity ends with students

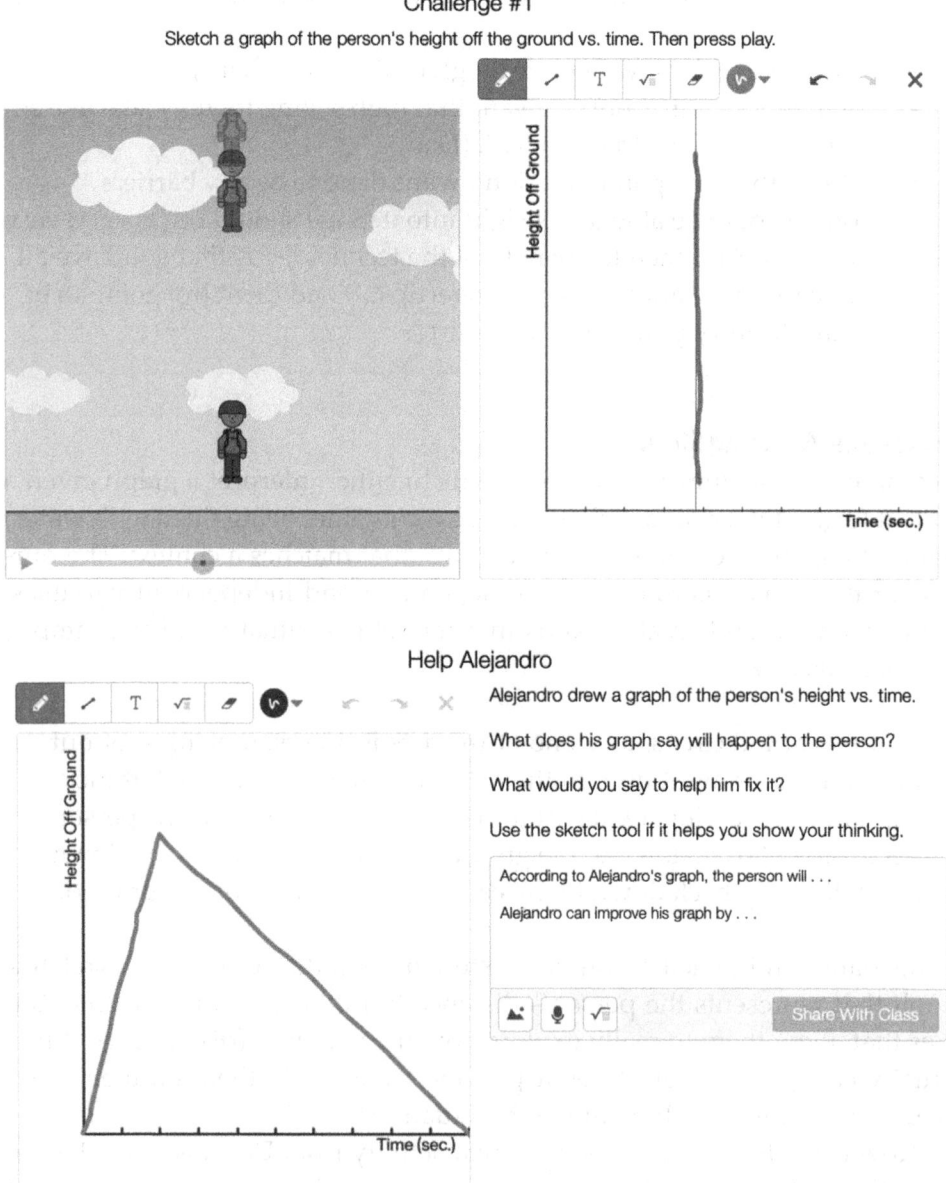

Figure 4.3 Cannon man from Desmos function carnival.

creating stories with matching graphs of their own. Using technologies like this provides a way for everyone to get started – just draw a graph! (low floor), and there are plenty of ways to explore (wide walls), as well as to be challenged (high ceiling).

Link 4.3 Desmos – Function Carnival

In the following sections of this chapter, we look at the specific characteristics of technology-enhanced tasks that lend themselves to being low-floor, high-ceiling, wide-walls types of experiences for students. Technology-enhanced tasks are made up of two parts: (1) technology sketches with which students can engage and (2) prompts or questions that guide students' exploration. In the case of Function Carnival, the connected representations – animation and graph – make up the "sketch", and the prompts include directions or questions such as "Sketch a graph of the person's height off the ground vs. time. Then press play" or "Alejandro drew a graph of the person's height vs. time. What does his graph say will happen to the person? What would you say to help him fix it? Use the sketch tool if it helps you show your thinking". We are separating out the prompts from the sketches as often prompts are posed separately from the math objects included in the technology sketch (e.g., on paper, on the board at the front of the room, or within an activity builder) and because many technology sketches can be used in different ways with different prompts.

Drawing on decades of design-based research (e.g., Desmos, 2016; Moynihan et al., 2021; Sinclair, 2003; Underwood et al., 2005), we first discuss the characteristics of the design of the technology sketches and then the characteristics of the types of prompts that ensure that when paired with the sketch(es) the resulting task positions students as math explorers, providing ways in which each and every student can explore while supporting their learning of the specific goals the task was designed to address.

Principles for Designing Technology Sketches for Exploring Mathematics

A high-quality sketch uses math action technologies in ways that allow students to engage playfully and make clear connections among the different representations included. Often described as "user-friendly", well-designed technology sketches allow students to explore, are mathematically correct, draw attention to important tools and representations, and are uncrowded. We describe each of these principles next.

Technology Must Allow for Exploration

If we are going to position students as math explorers, the tools provided must allow for exploration. That means that the technology must support experimentation. Students need to be able to both freely explore if they choose to do so, as well as test ideas and get feedback. Ultimately you want students to be able to create something that is worth talking about! The ways they can explore and use the various representations included in the sketch end up becoming a shared image for the class to consider and discuss.

- "It would be something that allows students to discover the properties of a shape through movement or measurement. Things that they can experiment on dynamically". (Michele)
- "I want them to be able to play around with things. To be able to explore I guess is the best way to say it". (Nathan)
- "It needs to be interactive". (Leah)

Listening to the Tech-Math Teachers, we get a sense of the importance of making sure that students have the tools they need to explore and that representations are dynamic so students can interact with them. In addition, whether students are exploring freely or testing specific ideas, the sketch needs to be designed in such a way that they get feedback to prompt them to continue to think about and revise their thinking.

- "Whatever students are making in GeoGebra, if they think of a transformation and it doesn't show up as what they think. That's their feedback right there". (Leah)
- "What goes along with that is the feedback the students are getting through the technology is less evaluative. It's not whether you're right or wrong. It's that interpretive feedback. It's interpreting mathematically like, 'Hey you said [the] scale factor is three. This is what it's going to look like. You said the scale factor is negative two. This is what it's going to look like mathematically'. So, it's not necessarily right or wrong and there'[re] lots of different ways in the beginning to be right or wrong as well". (Nolan)

Note that the Tech-Math Teachers are thinking about whether the technology sketch provides students with *interpretive feedback*, meaning that the technology produces something in response to the student's action that provides them an opportunity to reflect on their action. Being able to see the results of their actions on mathematical objects is necessary to support experimentation. The feedback should prompt more thinking! It isn't whether the outcome is right or wrong, but how is the outcome related to the action you took? Why? Is it what you expected? If not, why not?

- "I really don't like when students are using technology and they just get a check mark or an x mark when they try to put something in. To me the check mark really does a lot of hurt sometimes. Nobody likes seeing that red check mark. I prefer trying to make interpretive feedback where they input something to see how it changes the graph, or how it changes the size of a polygon. That's the key thing that's really important to me, is that if they're using technology it's not just black-and-white. It's more of 'Hey, this is what you did wrong, but now here's something else. Here's another aspect of it, can you see why your thought was wrong and now can you adapt'". (Nick)
- "It's when a kid puts in something just random you know. Or what's the equation of this line? And then you know they type an equation of the line, and then that line appears on the graph. So they have some feedback off of whatever their response is. There's this dynamic relationship going back and forth of not, 'Oh, you got it wrong'. But well, this is what your graph is. Do you want to go back and fix it? Or did you get it right? You know, like it gives them some informal feedback without the pressure". (Lauren)

What this means as a design principle is that we need to think about what tools and representations we can make available to students so that they can learn about their actions and responses from exploring with the technology. Ultimately the hope is that through their explorations, students create things worth talking about.

- "It needs to allow students to see lots of different situations all at the same time, or give students a chance to pose questions and think about – okay why does this work?" (Nathan)
- "Can they compare and contrast the things that they're seeing either in small groups or whole class where they can talk about the mathematics in some sort of way to their peers?" (Kristen)

An Example: Triangle Buster

A common conception that students have about geometric figures is that they can make assumptions about the figure based on the way the image "looks" on the page. For example, when given a triangle with a segment that connects a vertex to the opposite side, it is common to assume that the segment bisects the angle and the side. One way to support students' refinements of their conceptions about triangles is to pose questions using both static and dynamic images. When talking about the importance of a task positioning students as explorers and providing them feedback worth talking about, Karen shared this task as an example.

Students are first asked questions about triangles using static images like those shown in Figure 4.4. Notice that the triangles look the same but have different conditions.

Link 4.4 Full Triangle Buster Handout

Circle if the conditions are true for each number **Always, Sometimes,** or **Never.**

1. $\overline{AD}$ bisects $\angle CAB$. What do you know?

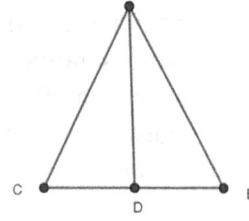

    a. $\angle CAD \cong \angle BAD$ A   S   N

    b. $\overline{CD} \cong \overline{DB}$ A   S   N

    c. $\overline{AD} \perp \overline{CB}$ A   S   N

    d. $\overline{AD}$ bisects $\overline{CB}$ A   S   N

    e. $\angle CDA$ and $\angle BDA$ are right angles   A   S   N

2. $\overline{AD}$ bisects $\overline{CB}$. What do you know?

    a. $\angle CAD \cong \angle BAD$ A   S   N

    b. $\overline{CD} \cong \overline{DB}$ A   S   N

    c. $\overline{AD} \perp \overline{CB}$ A   S   N

    d. $\overline{AD}$ bisects $\overline{CB}$ A   S   N

    e. $\angle CDA$ and $\angle BDA$ are right angles   A   S   N

Figure 4.4 Triangle buster handout.

PAUSE AND CONSIDER

1. How do you anticipate students would respond to the questions in Figure 4.4? What reasons might they provide for their responses?
2. In what ways would a preconstructed sketch based on the given conditions (a sketch for Question 1 and another for Question 2) support student reasoning?
3. What would you want to watch out for to ensure such a sketch would not add to students' confusion?

After completing the paper-and-pencil questions, students are prompted to open a GeoGebra sketch in which they see the same questions but have tools available to explore and test their ideas (Link 4.5). By adding measures and dragging the vertices of the triangles, students get interpretive feedback to inform their ideas about the relationships between the given conditions and what else might be true. As they drag and create different triangles, they are creating images worth discussing as they will support their conjectures and provide examples to share with others. For example, in Figure 4.5 on the left, students have added angle measures and see that the angles are congruent as expected given the original statement. They might assume the triangle is isosceles based on how it looks. On the left, the students have not only added additional angle measures but have also dragged vertex C down and away from point D. Now they can see that the triangle is definitely not isosceles. The tools provided within the sketch were carefully selected so that students would have the tools they need but wouldn't be overwhelmed by the tools that would not be helpful. The carefully constructed GeoGebra sketch provided an easy way for students to quickly and accurately explore properties and test their conjectures.

Link 4.5 Triangle Buster Exploration GeoGebra Sketch

Draw Attention to Important Tools and Representations

For a sketch to be user-friendly it is important that tools and representations are easy to see and talk about. For example, in the GeoGebra activity that Karen created to explore triangle conditions, the tools available to students were clearly lined up at the top and the vertices that could be dragged within the sketch were a different color from those that could not. However, as sketches get more complex (e.g., complex geometric constructions or multiple graphs, tables, and equations), the use of color and other markings help to distinguish between representations of the same type and to show links

1. Segment AD bisects ∠ BAD. What do you know?

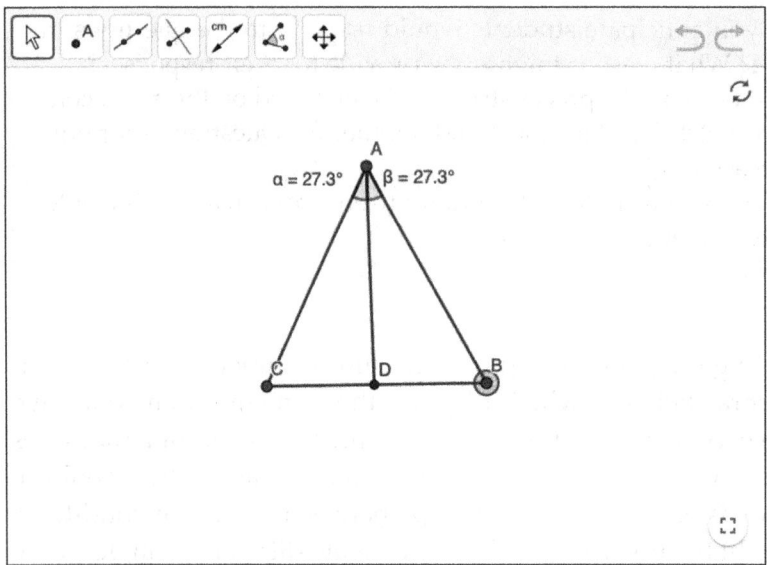

1. Segment AD bisects ∠ BAD. What do you know?

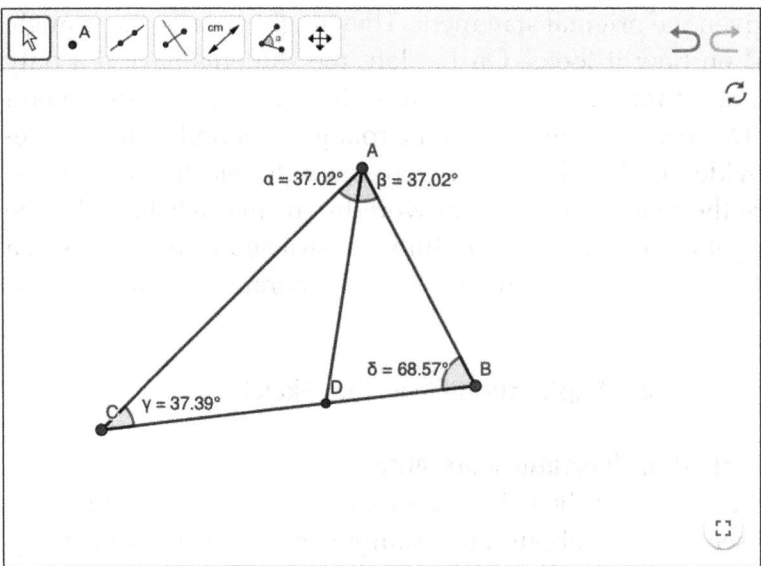

Figure 4.5 Student-created examples using the triangle buster GeoGebra sketch.

between connected representations. The Tech-Math Teachers indicated that the use of color and other markings to help distinguish or connect an important design principle.

- "Color and stuff … I do that a lot of times to try to point out similarities or different aspects that I want to stand out as they're looking at things". (Michele)
- "I try to distinguish different colored graphs and different colored features within like an image itself. … So, when I am looking at a task if they have two graphs and they're both black lines or something we can't really tell that they're supposed to be two separate lines or maybe is one an absolute value function. I don't know what's going on here. So I always tend to edit and make the colors different". (Samantha)
- "I guess within the graphing side of things I do want there to be color in there because otherwise I feel like it's really easy to lose which function is which, especially if you're trying to compare multiple things". (Nathan)

In the graphs shown in Figures 4.6 through 4.8, color and different styles (e.g., solid, dashed, dotted curves) are used to distinguish between functions. Notice that in all the examples, color is used to connect graphical and symbolic representations. However, in each of the examples, the symbolic representations are both placed and linked differently. In the first graph (Figure 4.6), the symbolic representations appear on the left only, and students would need to read the sliders to determine the specific function

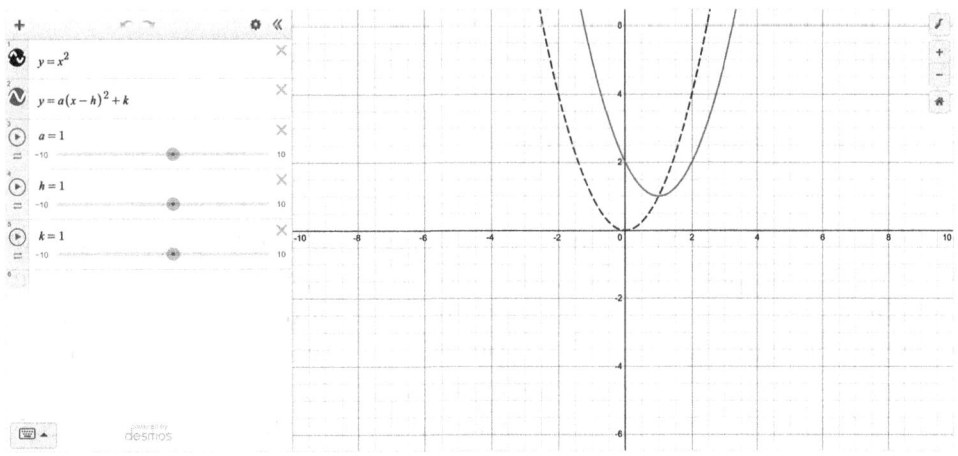

Figure 4.6 Sliders on the left.

Drag the slider k to change the parameter value in the red function, y = 1(x - 1)^2+-3, to explore the effect on the parent function (shown in black).

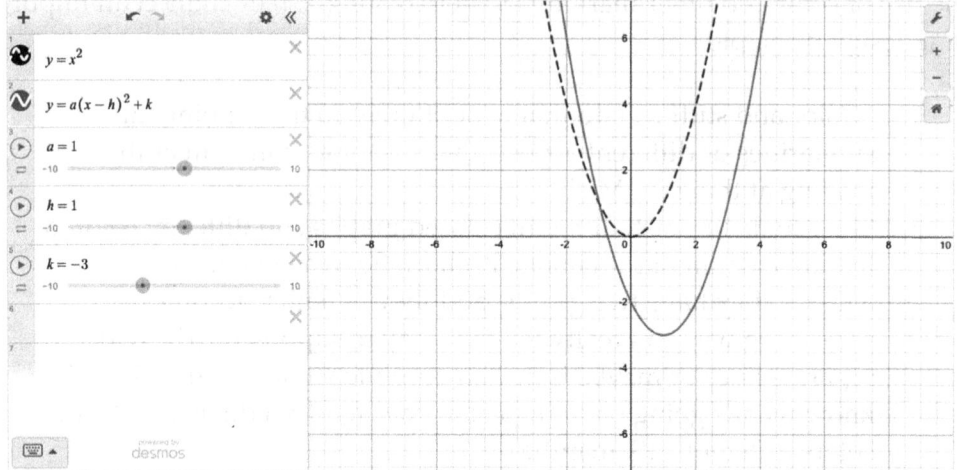

Figure 4.7 Dynamically linked function in the heading.

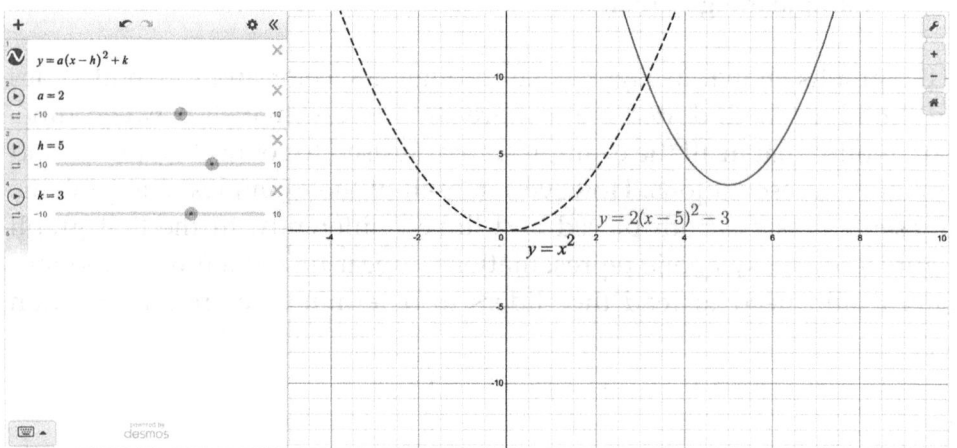

Figure 4.8 Dynamically linked function in the sketch.

represented in the graph. In the second (Figure 4.7), linked text is added to the top of the page. As the sliders are changed, the values within the symbolic notation change accordingly (this is sometimes called "hot text"). While the linked text is included, it is placed in an area of the page that many students overlook, so they might not even see that it is included. In the third example (Figure 4.8), linked text is included in the cartesian plane and colored to match the graph it is connected to. When placed within the plane, it is very difficult to overlook!

Using color and markings not only make a sketch easier to understand but also make it easier to talk about. When moving from informal to formal language, color provides students with a way to easily describe what they are observing as they interact with a sketch. For example, in the graphs referenced above, students might compare the red to the black function (e.g., "the red function is two left of the black function and up three") and later talk about the black function as the parent function and the red function as the transformed function (e.g., "the vertex of the transformed quadratic function is two left of the parent function, and up three").

- "They feel more comfortable at first talking about the red function and the blue function of the purple function versus jumping in to say $g(x)$ came from $f(x)$. [Referring to comparing two composite functions] so I think the use of color has been really nice". (Kristy)

While the use of color and markings is definitely helpful, it is important to consider the students in your class when selecting the color and markings. Not all students can distinguish between colors, so it is important to be aware of that when choosing the colors in any sketch. (And when thinking about the ways your projector might additionally distort the colors, making them difficult to see.)

- "If I'm talking about the red graph I usually try to put that text in red, but I'm also trying to remember if someone cannot see red or cannot differentiate between red and green that I need to have some other ways. So that's why we have dotted lines that we have solid lines". (Andrew)

PAUSE AND CONSIDER

In Figure 4.9, is a sketch designed to support student exploration of the relationship between the radius and the height of a cylinder and its volume (Link 4.6). The radius and the height can be changed by dragging the sliders and the cylinder construction, base measure, and volume measure change accordingly. How might you use color or markings to draw students' attention to important connections?

Link 4.6 Volume of a Cylinder Sketch

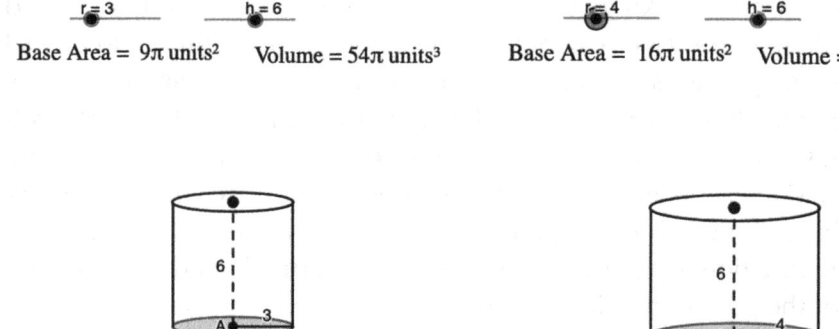

Figure 4.9 Volume of a cylinder sketch.

One way to draw attention to important relationships is to use color to match the slider to the part of the figure it is controlling. For example, I could make the r slider red and then the radius marking and the measure of the radius red as well. The same could be done with the h slider, the height marking in the sketch, and the height measure. The base area measure could be colored the same as the shaded base area in the figure. Finally, the volume measure could be the same color as the entire outline of the figure indicating the three-dimensional volume represented. However, to do this would require four different colors, so you would want to choose those colors carefully. It is important that they are easy to distinguish from one another for all students (including any students that might see color differently).

Make Sure the Sketch is Mathematically Correct

Luckily the technologies typically used in secondary classrooms today are mathematically sound, meaning that they accurately react to students' actions in mathematically defined ways. However, if you are using a preconstructed sketch, it is important to check that all aspects included in the sketch are mathematically correct – that includes checking any notation, vocabulary, mathematical relationship, or explanation.

- "A good technology-based task for me doesn't have too many mistakes in it. I have made plenty of tasks and they have gone not so smoothly because I have made a lot of errors. So I always look for that there aren't errors, specifically math errors. That's what really actually throws the kids off". (Samantha)

For example, if you are hoping for students to explore the properties of a rhombus it would be important that the sketch includes a rhombus that can be dragged and remains a rhombus (i.e., it passes the drag test). In addition, it would be important that the sketch when dragged results in non-square rhombi as well as square rhombi. So, while proofreading all text included in a sketch is important, checking all the dynamic mathematical or statistical objects included is as well. You might even ask a friend to try to break it for you.

Space is Used Efficiently

While color and markings can help draw students' attention to tools and relationships, if a sketch is too crowded or too wordy, even the best use of color and markings is not going to help students navigate it.

- "I want to be as clean as possible and where the student can read, they just need to read a little bit". (Dan)
- "In the physical design, some of the things that are really important to me. So, like if it's too complex or it's too muddled, like if I'm comparing six things on a screen that might make things too difficult for students to focus on the thing that I really want them to uncover. Putting too many things in the space is a problem". (Kristen)
- "It needs to be something that is easy for students to read and understand. I've done it myself where I write too many words on a screen and I lose them". (Leah)

> **PAUSE AND CONSIDER**
>
> In Figure 4.10, is a preconstructed sketch intended for students to explore the relationship between chords that are equidistant from the center of a circle and the measures of their minor arcs and central angles. What changes would you make to this sketch so that it is less crowded?

One way to make the space less crowded is to use checkboxes or buttons to turn parts of the sketch on and off (i.e., hide and show). This would allow students to choose the objects and measures they want to focus on as they explore. Another strategy is to move the questions to a different location. Sometimes it is more helpful to have the questions separate from the sketch. This can be done by putting them into an activity builder or on a worksheet. Figure 4.11 shows the same sketch (created by Samantha) using checkboxes to make the space less crowded and color and markings to highlight important connections. The instructions and prompts were moved to a worksheet for students to record their ideas on as they explored using the sketch.

Link 4.7 Samantha's Circle Chord Relationships Task

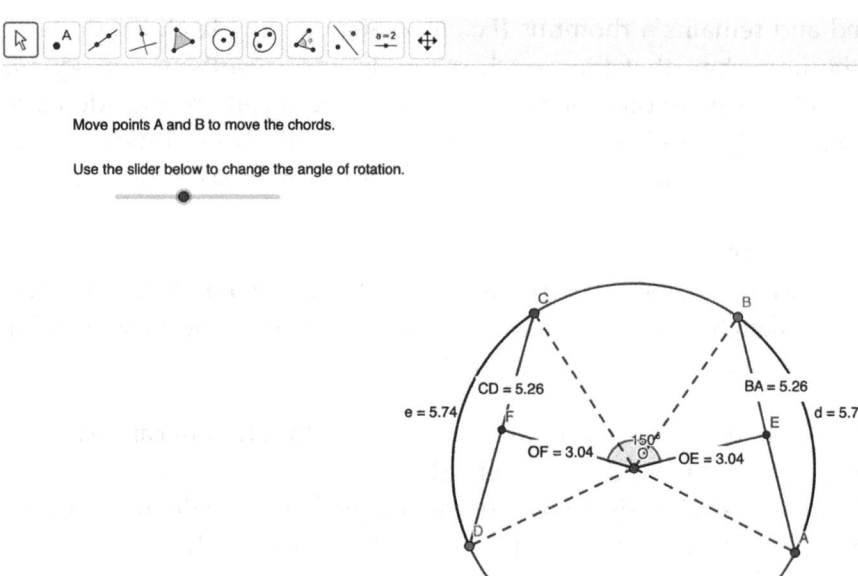

1. Describe what geometric objects are shown.
2. Explore and note three things that you notice and one that you wonder about.
3. What conjectures can we make about chords that are equivalent?
4. What is the relationship between triangle BOE and triangle DOF?

Figure 4.10 Circle chord relationships preconstructed sketch 1.

Prompts to Accompany Technology Sketches for Exploring Mathematics

Just as important as the technology students are engaging with are the prompts we provide or questions we pose to guide their exploration. If we are going to position students as math explorers, then the prompts we use with technology-enhanced tasks should encourage exploration. Some sketches you find might have prompts already included, other times you might find a great sketch and need to create the prompts to accompany it yourself. Either way, the design of the prompts is very important to consider.

The Tech-Math Teachers pointed out that in examining prompts, they are looking for questions that promote thinking and that will ideally give students something interesting to talk about. One of the easiest ways to do this is to ask for informal ideas before formal ideas. As noted earlier, if the sketch is set up so that students can easily explore and get interpretive feedback, then including prompts that ask them to do exactly that is a great way to get started. You can then ask them to build on their informal ideas in more formal ways.

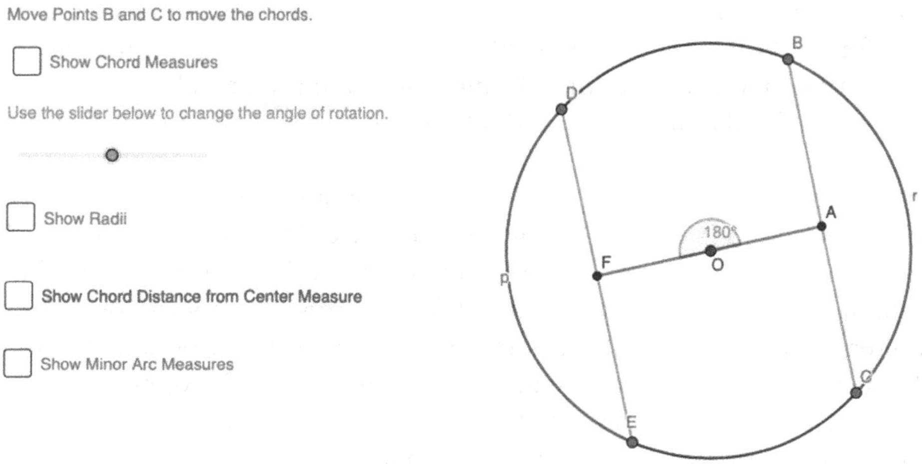

Move Points B and C to move the chords.

☐ Show Chord Measures

Use the slider below to change the angle of rotation.

———————◯———————

☐ Show Radii

☐ Show Chord Distance from Center Measure

☐ Show Minor Arc Measures

Investigation 2:
For this investigation use Page 2: Investigating Equivalent Chords, in the Investigating Chords
GeoGebra book.

1. Describe what geometric objects are on the page. Explore and note three things that
    you notice and one that you wonder about.

2. Only show the measures of the minor arcs and the measures of the chords. Drag
    points B and C and write down your observations.

3. What conjectures can we make about chords that are equivalent? Based on your
    observations from question 2.

4. What is the relationship between triangle AOB and FOE? Explain.

5. Justify your conjecture in question 3.

Figure 4.11 Circle chord relationships preconstructed sketch 2.

- "I think particularly with technology you want to give them a way
  to talk about what they're seeing in any formal or informal language.
  So they may look at something, may not know what it is, but they
  can talk about what patterns they pick up on and it gives all students
  the ability to do that". (Kristen)
- "I want them to have that sense of exploration and play, and I want
  it to build from informal to formal". (Lauren)
- "I think about how do I invite informal observations before the
  formal. So kind of low-floor, high-ceiling kind of questions where
  you don't have to know any math to answer the question. So a lot of
  the questions are those notice and wonder questions or just what do

you see going on. Those types of prompts that are informal before formal. Something where students are able to interact and maybe start to make those informal connections themselves, also ways to keep it kind of low stakes". (Nolan)

An easy way to promote exploration and sharing of informal ideas on which you can later build is to ask, "What do you notice? What do you wonder?" Asking what students notice and wonder is powerful because everyone has something they can notice (Ray-Riek, 2013)! Prompts like this are low stakes and often bring out brilliant ideas that you can later build on and connect to the formal goals of your lesson. After students have the opportunity to explore and express their brilliant ideas in their own words, then you can start to bring in prompts that are more tightly tied to your goals yet still encourage exploration. A collection of prompts along this line encourages autonomy for making and testing conjectures, examining outcomes, describing them in a way that makes sense to oneself, and then building on those ideas connecting to the formal mathematical language, notation, concepts, and procedures.

When looking over a collection of prompts, consider not only if they move from informal to formal but also if there is variety in what is being asked. No one wants to answer the same question over and over. In addition, if there is a particular representation that you hope students pay attention to, it is important to ask them to do so. The curriculum designers at Desmos (2016) explain,

> An activity becomes tedious if students do the same kind of verb over and over again (calculating, let's say) and that verb results in the same kind of noun over and over again (a multiple choice response, let's say). So attend to the verbs you're assigning to students. Is there a variety? Are they calculating, but also arguing, predicting, validating, comparing, etc? And attend to the kinds of nouns those verbs produce. Are students producing numbers, but also representing those numbers on a number line and writing sentences about those numbers?
>
> (para. 4)

The Tech-Math Teachers note that they are guided by this principle as well.

- "I try to write questions that are varied and ask maybe something from a different direction". (Michele)
- "I try to make sure that there's a variety of prompts. I try to start in a way that's inviting and then have my prompts kind of get more expansive as we go". (Kristy)

Advice for Selecting Technology-Enhanced Tasks

While we hope that it is clear that selecting good technology-enhanced tasks is not trivial, we don't want it to feel too overwhelming. Rather, keep in mind that there are great tasks out there already, so knowing what makes a good one will make sifting through them much easier. If you are just getting started with selecting (or adapting) technology-enhanced tasks, the Tech-Math Teachers provide four solid pieces of advice.

Start Small

While we are definitely arguing that there are concepts that are better taught using technology, we also fully recognize that overhauling your entire course at once is not realistic! Unless you are working with a curriculum that has quality technology-enhanced tasks embedded, then an important piece of advice is to start small!

- "It takes a long time to get into all this. Choose one piece of technology and maybe choose one or two lessons that you want to do with it. So maybe if you want to try to do a Desmos activity for the first time, find an activity someone else has already made that'll fit in your classroom and start with that. Then maybe the next time you do the Desmos activity maybe try a screenshot feature and see how that works with one or two screenshots. So very much start small and then maybe down the road they start trying to build a simple activity of your own". (Leah)
- "Don't feel like you have to use technology every single day in the classroom. I would say that because you're not always going to find exactly what you want, and if it doesn't work out use another resource". (Nathan)
- "Don't feel overwhelmed. Just first start with just finding one technology that you want to start using. I think it's very easy to hear all about these technologies. Especially like scrolling through Twitter or seeing what your colleagues are doing. I know I felt very overwhelmed at first". (Kristy)

Don't Reinvent the Wheel

Between Google Search and specific technology search functions (e.g., Geo-Gebra Resources, Teacher Desmos, Mathigon, NCTM Illuminations, Schodor), you have millions of technology-enhanced tasks at your fingertips. Often you can find something that fits your needs exactly. However, when you can't find exactly what you are looking for, find something close and use the design principles outlined in this chapter to adapt it to make it into exactly what you need!

- "GeoGebra has a ton of tasks already created. Desmos has a ton of tasks already created. So, explore those resources that they have first and then if you don't love what you see find a way to modify it. But I would try to find what's already created". (Nathan)
- "Start with a predesigned task. Just thinking about Desmos, they've already got some curated collections that already have teacher tips in them. So implement one of those first". (Kristen)
- "Don't try to reinvent everything, just look for what's out there. Know what point you want to get across, know what concept you want to get the kids to understand, look for what's out there and just modify things that are already there. Don't try to build everything from scratch because that is just way too time-consuming and you will get lost". (Andy)
- "I really don't encourage most people to create their own thing. I really encourage them to modify. Find something you like, and if you want to change something go ahead and modify it". (Dan)

Engage with the Task

Before using a technology-enhanced task with your students, it is important to engage with it yourself. It is through engagement that you pick up on ways that the sketch and associated prompts can be improved. In addition, through engagement, you'll have better insight into the various ways students might engage with the technology and strategies they might use to address the prompts. Thinking about your students' strategies and how what they will see as a result of the actions they take will help you anticipate their thinking and plan assessing and advancing moves.

- "You've gotta experience it as a teacher first to think about where the hiccups are. It's planning, that's one of the things with technology, I think you've really got to plan ahead and experience it yourself. Find your little typos or that interactivity that you're talking about, 'Hey I designed it wrong'. When I'm designing my own activities as I finish each screen I do it and I just and make sure. Just to make sure my students are having a positive learning experience. It basically comes down to preplanning, making notes as you're going along, then your reflection afterwards". (Nick)
- "Make sure that if you select a task that you work through it at least once on your own, if not multiple times to try to see multiple approaches for that task. Or maybe things that you have seen students do". (Nathan)

Collaborate with Other Tech-Using Teachers

Having a professional learning community is essential to feeling supported when trying out new things. So find people within your school, district, at conferences, or in an online community to collaborate with! On social media (e.g., Twitter, Facebook, Instagram), there are groups of teachers sharing technology-enhanced tasks, helping each other troubleshoot, and generally supporting each other in trying out new ideas. (The hashtags #mtbos and #iteachmath are great ones to follow.)

- "Collaborate with teachers, whether it's teachers in your own school you know, make connections. The more I went to conferences, the more connections I made. With all the technology we have to collaborate, we don't have to be just with the teachers in our own building. So, create that support network for teachers that you can send 'Hey what do you think of this activity' and they can run through it". (Nolan)
- "Twitter is a fantastic place for this. There's even a hashtag, #improvemyAB (for 'improve my activity builder') that people will go in and just say 'This isn't working, can you help me make this connection work?' Or 'I'm really looking for something on linear functions and how to really show slope. Do you have anything?' The math teacher community is amazing! Use it and start small and you'll build the skills as you go". (Andrew)
- "There are things like help forums out there as well that you can go to and like ask questions. … People are usually so quick!" (Allyson)
- "Some of my best activities I've made myself are on their third and fourth iteration. I get feedback from other teachers, like 'Hey when you taught this, what's something that went well for you? What's something that didn't go well?' So, we can sort of work together. You know teachers need to help each other out". (Nick)

Keeping the Tech-Math Teachers' advice in mind will help you from becoming too overwhelmed by selecting a task aligned with your lesson goals.

CHAPTER TAKEAWAYS

To position students as math explorers, technology-enhanced tasks should have a low floor (everyone has a way to get started), a high ceiling (there are lots of possibilities for taking things even further), and wide walls (there are multiple pathways to a solution). This is accomplished using the following design principles for both the technology sketch(es) and their accompanying prompts – remember that the sketch and prompts work together to ensure a technology-enhanced task is of high quality and that it creates situations worth talking about.

Technology Sketch		Accompanying Prompts
<ul><li>Allows for exploration</li><li>Draws attention to important tools and representations</li><li>Represents the math correctly</li><li>Space is used effectively</li></ul>		<ul><li>Are clearly and succinctly worded</li><li>Move from informal to formal</li><li>Use a variety of verbs (actions) and nouns (representations)</li></ul>

Questions to Discuss With Your Colleagues

1. Certainly there are additional principles for selecting or designing technology-enhanced tasks that could be added to the preceding list. What would you add? Why?

2. Do a quick search for a technology-enhanced task for a topic you will be teaching sometime soon. Open the first task you find and critique it with respect to the design principles outlined in this chapter. Be sure to consider both the technology sketch and the accompanying prompts in your critique.

3. Implicit in the design principles presented in this chapter is a vision of teaching math with technology that values students as individuals with brilliant ideas that should be highlighted and built on. What are some sample prompts that could be included with the task you found in Question 2 that would align with this vision? Explain.

 You can find links to all the technology-enhanced tasks and supplementary videos throughout the book at https://www.tlmtresearch.com/teachingmathtechbook.

5

Launching Technology-Enhanced Tasks

The way we set up a task (introduce it to students) is sometimes referred to as *launching* a task (Lappen et al., 2009). The way a task is launched impacts both what you and your students are able to achieve during the lesson. We've all set students off to start on a task only to see hands popping up all over the classroom. They all have questions about how to get started. Running around putting out fires or stopping everyone to give more instructions is a sign of an ineffective task launch. That is what we are trying to avoid. It wastes our precious time together, and when you make adjustments in the moment, it is so easy to accidentally give the key idea of the task away, robbing students of the opportunity to think. An effective task launch is one in which all students have what they need to get started and the cognitive demand has stayed the same as originally intended. Here are a few things that the Tech-Math Teachers noted about launching:

- "I feel like when it's technology-based usually it's more exploratory so it [the launch] needs to be even more clear. So, when it comes to the launch the objectives are super important". (Nathan)
- "I think the main difference that I consider when I'm launching a technology-enhanced task is drawing students' attention to the features of what you're hoping they're looking at. If it's their first time using an applet or technology of some sort … and there's specific things you want them to pay attention to, it's really critical to go through and say okay so these are the things that are dynamic

DOI: 10.4324/9781003302285-6

… pointing out the things that are dynamic and what you want them to pay attention to. So, like let's say I give them a table or graph and equation, do I want them paying attention to all three? I'll point out where all three are so that they're considering all three". (Kristen)

- "We sit for a second. I put up the first slide, or the first whatever that I want them to look at, and I model for them a little bit of how it works. Here's the button to make it full screen. If you don't want to be in full screen anymore you hit escape. If you click and drag on parts of this picture it'll move it around. You can click and drag in the corner and move the shape. If you zoom in and zoom out it will zoom in and zoom out. So just very, very basic things. You don't just want to give it to them and be like okay go, because you'll have kids raising their hand in a few minutes and they're like I made my shape really small and I don't know how to fix it". (Samantha)

- "The biggest one for me, Is can everybody do it? So, for me it's super important to make sure that that launch engages every single student right off the bat". (Zachary)

The Tech-Math Teachers noted important aspects of a launch that are aligned with what research has shown to be effective. In fact, based on research in more than 150 classrooms, researchers have identified four key features of an effective task launch – discussing the key contextual and mathematical features of the task, developing a common language to describe those key features, and (very important) maintaining the cognitive demand of the task (Jackson et al., 2012). When using a technology-enhanced task the same elements stand, but they might look slightly different.

Characteristics of an Effective Launch of a Technology-Enhanced Task

CHARACTERISTICS OF AN EFFECTIVE LAUNCH OF A TECHNOLOGY-ENHANCED TASK

1. **Discuss the Key Contextual Features of the Task**
   If the task is situated within a context, some students might have trouble getting started because the context or scenario is unfamiliar. Thus, it is important to discuss any features of the context of the task

that might be unfamiliar. You might do this through sharing pictures or video, asking students to imagine the situations, using digital simulations, or making connections to people, places, or things that you think might be more familiar to them.

2. **Discuss the Key Mathematical Ideas of the Task**
Being able to engage in the task means being able to interpret key mathematical ideas presented in the task. This includes addressing basic barriers regarding language and ensuring that students have an image of the mathematics represented in the task. If the task includes dynamic representations of the key mathematical ideas, it is important that the launch include ways to interact with the technological representations to investigate the key mathematical ideas.

3. **Develop Common Language to Describe Key Features**
Effective launches are those in which you don't just talk to the students but engage them in the conversation so that a common language is developed when identifying the key features (contextual and mathematical) of the task that are central to students successfully beginning the task.

4. **Maintain the Cognitive Demand**
Throughout the launch, it is important that the cognitive demand of the task is not lessened. For example, when discussing the key mathematical ideas, it is very important not to suggest methods to solve the task. Doing so robs students of the opportunity to develop important understandings and practices.

(adapted from Jackson et al., 2012)

PAUSE AND CONSIDER

Consider the Nets of Cubes task shown in Figure 5.1. The prompt asks if any arrangement of six squares can fold up to be a cube.

- What would you want to make sure you included in a launch of the task? Explain.
- What would you want to make sure you did not include? Why?

A cube has 6 square faces.
Does that mean that any arrangement of six squares can fold up to be a cube?

Figure 5.1 Nets of cubes task.

Link 5.1 Nets of Cubes Task

When thinking about the launch it is helpful to first think about the context, in this case, cubes. So, it will be important to begin the discussion of this task by talking about cubes. You might show a cube (maybe even hand each student a small cube) and ask what all cubes have in common. It will be important that we all agree that all cubes have six square faces. This is a good opportunity to develop a common language as well. Some students might refer to the sides of the cube while others refer to the faces of the cube. Take a moment to talk about these two terms – sides versus faces: Are they the same? Which should we use? Why? Next it is important to consider the key mathematical ideas – in this case the key idea is a net of a cube. It is important that students understand they are trying to see if each of the arrangements of squares can be folded up to make a cube (i.e., is a net of a cube). Part of determining whether or not the arrangement can be folded into a cube is being able to use the Polypad tools to fold and unfold the net. You might start with a basic arrangement (1 × 6 or 2 × 3) and ask if it will work. Change the color of each square so it is

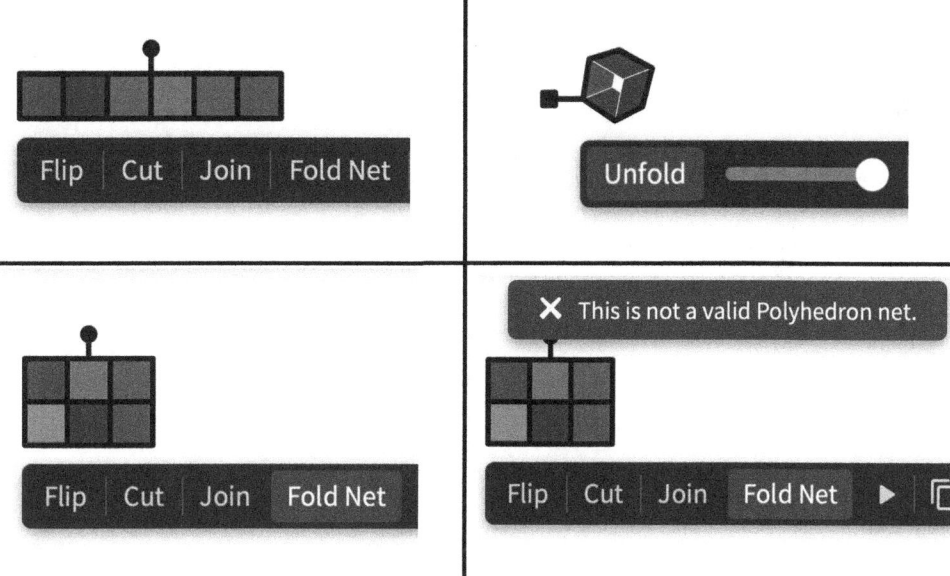

Figure 5.2 Demonstration of the polypad tools during a launch of the task.

easy to see when folded, and then show students how to select all the squares in the arrangement and select the "fold" option (Figure 5.2). Does it work? Make sure students do this with you at least once so they are all familiar with the tools. Finally, it is important *not* to reduce the cognitive demand of the task, so while the 2 × 3 arrangement does not work, don't say why, just "Hmmm, I wonder why that one doesn't work?" Make clear to the students that figuring out why some arrangements work and others don't is what they are working to figure out. At this point, the students should be ready to go!

Examples of Effective Technology-Enhanced Task Launches

Next, we share two vignettes of technology-enhanced tasks that Kristen, one of our Tech-Math Teachers, launched in her classroom. Following each vignette, we discuss how Kristen included each component of an effective launch.

Example 1: Pixar vs. DreamWorks

There has been a years-long debate over which animation movie studio is the "best". DreamWorks or Pixar? Favorite movies from your students' childhoods come from these two companies – *Toy Story, Minions, Cars, Up, Shrek*. This is a nice context for an exploratory data analysis task. Students are provided data from all of the movies released by both

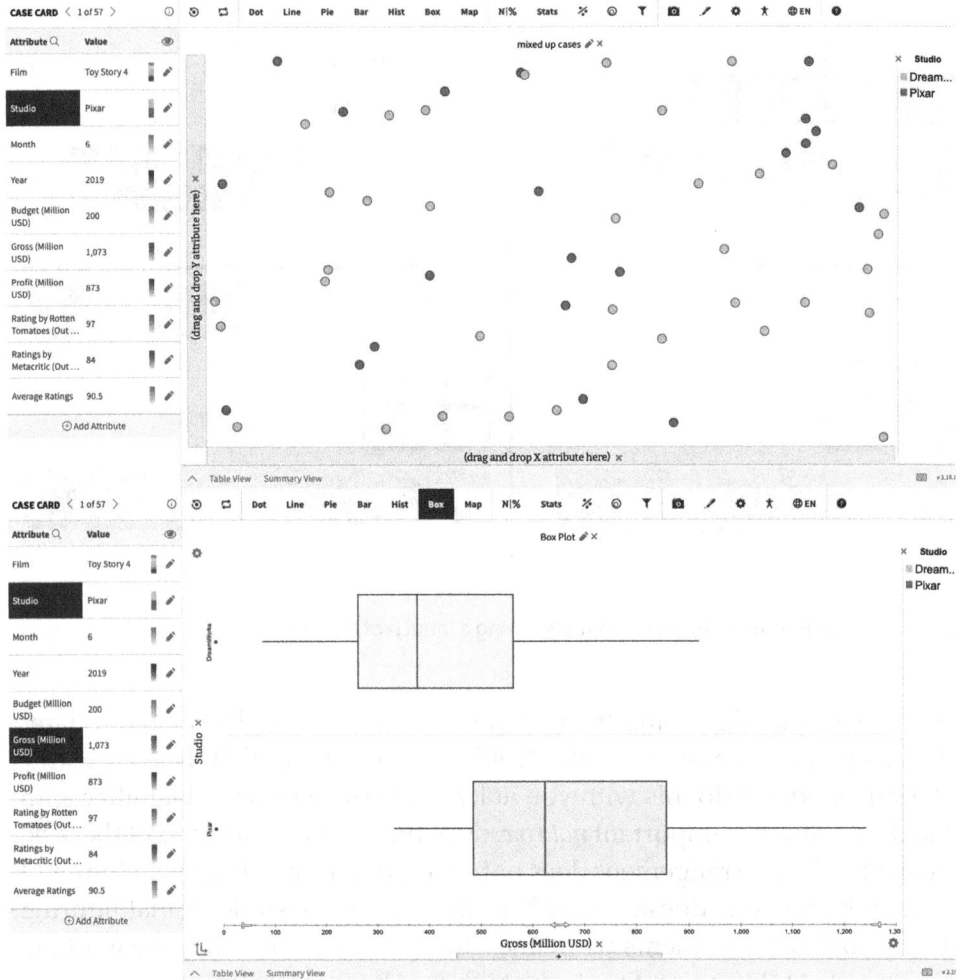

Figure 5.3 Pixar vs. DreamWorks data exploration using Tuva.

companies (e.g., date of release, budget, gross, profit, ratings) in a dynamic statistics application (Tuva is used here) and simply asked "Which company is more successful?" Then students explore the data, creating representations, statistics, and a data-based argument for their solution (Figure 5.3).

Link 5.2 Pixar vs. DreamWorks Data Exploration Using Tuva

PAUSE AND CONSIDER

Work through the Pixar vs. DreamWorks task. How would you launch this task? Consider all four elements of an effective technology-enhanced task launch.

Kristen invited us into her classroom to observe her teaching this lesson. Let's take a look at how she launched the Pixar vs. DreamWorks task. The transcript of the task launch follows. As you read, identify what you think are key aspects to the success of the launch.

Vignette: Kristen Launching the Pixar vs. DreamWorks Task

Kristen kicked off this task by asking the students, "What is your favorite animated Pixar or DreamWorks movie?" She provided them a place to record their favorites and went around the room to everyone who wanted to share quickly so they could say their favorite aloud. There was a lot of excitement (and disagreement)! She then played a short video compilation of Pixar and DreamWorks movies to help them begin to think about all the great movies that the two studios have released over the years (Link 5.3).

Link 5.3 Pixar and DreamWorks Epic Trailer Mash-Up

After the video was complete one student, Camron, stated loudly that Pixar was better than DreamWorks.

Link 5.4 Video Clip: Kristen's Launch of the Pixar vs. DreamWorks Task

Teacher (Kristen):	Okay. So, Cam kind of propositioned us towards the question that we're trying to answer today which is, which of these is better? DreamWorks or Pixar?	1 2 3
Students (chorus):	[Students saying their choice over each other]	4
Teacher (Kristen):	I'll hear out an answer or two and then I want to think about it for a second. Asch, do you want to start us off?	5 6
Asch:	Okay. Is *Bolt* a Pixar movie?	7
Teacher (Kristen):	I don't know.	8
Students (chorus):	Yes! Yes!	9
Teacher (Kristen):	All right, Jordan.	10
Jordan:	I think DreamWorks, but I do think Pixar is more con-sistent.	11 12
Teacher (Kristen):	Okay. Srikar?	13
Srikar:	DreamWorks, did make a lot of real good movies? Like *How to Train Your Dragon*? And what was that? Yeah, *Kung Fu Panda*. And Pixar also made movies like *Cars* and *Toy Story*. But since Pixar had more movies that I liked; I would go with Pixar.	14 15 16 17 18
Teacher (Kristen):	Okay. Cameron?	19

Cameron:	Okay. So, the way I compare the two, Pixar is like Disney and DreamWorks is like Nickelodeon. And I'm not gonna go into deep about that, because I don't even know what I'm saying.	20 21 22 23
Students (chorus):	[laughing]	24
Teacher (Kristen):	You brought up a nice metaphor.	25
Cameron:	DreamWorks, they have good movies on *How to Train Your Dragon*. *Shrek*, they have better movies and more better movies.	26 27 28
Teacher (Kristen):	Do you mean there are a bigger quantity of better movies?	29 30
Student (unknown):	Yeah.	31
Cameron:	They have *Cars*. *Toy Story*.	32
Teacher (Kristen):	All right, let's pause for a second. So, what I'm hearing a lot in this discussion, is this idea that we have some kind of connection. A lot of the arguments I'm hearing are, are very based on, our personal experience and what we like in a movie, like which ones we prefer? Right. Okay?	33 34 35 36 37 38
Student (unknown):	I didn't even know they were different things, like five minutes ago.	39 40
Teacher (Kristen):	Well, there we go. So there are two different studios that produce these animated films. And so a lot of the things you guys are telling me are very ingrained in bias. Does anybody know what I mean by that? Tsi, what do I mean by that?	41 42 43 44 45
Tsi:	You pretty much mean, like, we pretty much grew up with this one thing that we liked so much that we would rather say that this thing is better rather than the other thing.	46 47 48 49
Teacher (Kristen):	Right? So there's a lot of personal connection to the movies that you're watching. And so therefore you might be leaning a certain direction because of that, right? Okay. All right, Shumer.	50 51 52 53
Shumer:	It's really based on our opinion on what we like, so that's why it's biased.	54 55
Teacher (Kristen):	Okay, there's a big cloak of opinion here, right? Okay. What I want to do is try to answer this question without that cloak of opinion. Okay. All right. How can we answer this question without a cloak of opinion?	56 57 58 59

	So what might we be basing the criteria of how we're making a decision on? Ethan?	60 61
Ethan:	Movie ratings.	62
Teacher (Kristen):	Okay, so maybe I'm looking at movie ratings. Tsi, what do you think?	63 64
Tsi:	Like polls.	65
Teacher (Kristen):	Polls? So you mean like asking people what their favorite is and seeing which one is the most maybe? Okay. Jordan, what do you think?	66 67 68
Jordan:	Sales and revenue.	69
Student (unknown):	[can't hear]	70
Teacher (Kristen):	Sales and revenue. Okay. Kni?	71
Kni:	Okay, so the way that we do it is pretty much like so we can try and see which one had like better popularity and pretty much when they first released and then see how that moved on.	72 73 74 75
Teacher (Kristen):	How can you measure that popularity?	76
Kni:	By usually interviewing people or like yeah, pretty much everybody said.	77 78
Teacher (Kristen):	Eli?	79
Eli:	Box office	80
Teacher (Kristen):	When you say box office what do you mean?	81
Eli:	Like if like the like whichever one and like how high the box office is [unclear]	82 83
Teacher (Kristen):	That seems really tied to what Jordan was talking about. Right? Reade, what do you think?	84 85
Reade:	Most recognizable.	86
Teacher (Kristen):	Most recognizable so like, if more people can recognize it, then it must be a better movie?	87 88

At this point, Kristen directed the students to open up the link to the data in Tuva that she had posted on the board.

Teacher (Kristen):	This is called Tuva. Okay. We're going to talk about the features and make sure we understand what we're looking at here. Okay. And get a sense of how potentially we can use this to our advantage to maybe make a claim. A claim would be like which one is better? Right, Pixar or DreamWorks? And maybe provide some evidence? And then give some reasoning behind that evidence. Does that sound familiar? Anyone?	89 90 91 92 93 94 95 96

Students (chorus):	Yes.	97
Cameron:	CER.	98

Note: CER (Claim–Evidence–Reason) is a routine the students had been using in their English and social studies classes.

Teacher (Kristen):	Oh! CER. All right, here's what I want you to do first, everybody with me? So on the computer you guys are on, I want you to just close down this whole thing, hit that arrow. [Closing the right window.] So, it's out of our way that we do not need it. Okay, now our screens are bigger, and we can kind of play around more with the data that we're looking at. So, the first thing I want to point out is what we're seeing over on the left-hand side. What are we seeing? Anybody want to speak to it?	99 100 101 102 103 104 105 106 107
Students (chorus):	[naming things they see all at the same time.]	108
Teacher (Kristen):	Okay, [as she points to the left of the screen] so we see attributes, values. What are the attributes we have on this list?	108 109 109
Student (unknown):	Film, studio, month, year, budget rocks, I mean, gross, profit, rating by Rotten Tomatoes. Rating by Metacritic, average rating.	110 111 112
Teacher (Kristen):	Okay, so anybody see something on there that they're questioning? What are you questioning, Edwin?	113 114
Edwin:	What do the colors mean?	115
Teacher (Kristen):	There's some colors on there. So what do the colors mean? What's another thing we might be questioning? Jordan?	116 117 118
Jordan:	Who's Metacritic? Metacritic. Anybody know who Metacritic is? Okay.	119 120
Student (unknown):	Some random person online.	121
Teacher (Kristen):	I think the colors will become clearer here in a second. But let's answer Jordan. It's just another critic like Rotten Tomatoes. Srikar?	122 123 124
Srikar:	What is the value that is given for the attribute month?	125
Teacher (Kristen):	Hm, good question. What does this month mean? So we probably know what the year means. What do we think the year means?	126 127 128
Student (unknown):	The year means released.	129
Teacher (Kristen):	So when in the year was this released?	130
Student (unknown):	2019.	131

Teacher (Kristen):	When in the year, though?	133
Students (chorus):	June.	134
Teacher (Kristen):	And how are you figuring that out?	135
Students (chorus):	It's the sixth month.	136
Teacher (Kristen):	Okay, so that's what the month means, if I see the month. Okay, are there any words that we see over here on the left that maybe we think that either we don't know or maybe someone in the room might not know? What do we think? Say that again. Kni?	137 138 139 140 141
Kni:	What is gross?	142
Teacher (Kristen):	Gross? Do we know what gross means? What do you think, Jeremiah?	143 144
Jeremiah:	How much money they made.	145
Teacher (Kristen):	How much money they made? Can you elaborate a little bit on that?	146 147
Jeremiah:	How much money they made by ticket sales and all that.	148 149
Teacher (Kristen):	So, how does it differ from the other thing that we see up here which is profit?	150 151
Jeremiah:	Because it takes money to make the movie. And they take the money from their gross, what their budget was and they take that away.	152 153 154
Teacher (Kristen):	Oh, can you guys see that subtraction potentially happening here? Jeremiah is saying, if I take the gross and I take out the money that they put in the budget, that we end up with that profit number. So if we're looking at gross, gross would be the amount that the total amount of money I'm taking in, before I pay for like, my director, my budget, my voice actors, right, all the stuff that goes into it. Okay. Anything else over on the left-hand side, we think we need to clarify before we start moving some stuff around and see what was this thing can really do. Okay. All right. So let's take a look over here. I want to show point out a few things. So I'm going to show you some features. What is we're what are we mostly trying to compare in this scenario out of these attributes.	155 156 157 158 159 160 161 162 163 164 165 166 167 168 169
Student (unknown):	Is Pixar better than DreamWorks?	170
Teacher (Kristen):	And why? Right? Notice that some of the things you guys suggested like ratings and sales are showing up in these attributes, right. So I want to show you a little bit	171 172 173

	about how to work with this and all the features that are	174
	on it. Let's say first of all, if I click film, notice what hap-	175
	pens. It starts to highlight "film" over here. And now all	176
	the films have their own color, even though they're hard	177
	to distinguish because there's 57 on the list. All right.	178
	Might not be the most helpful thing to us. If we're try-	179
	ing to make a determination between Pixar and Dream-	180
	Works Which one do you think we should focus on?	181
Student (unknown):	Studio?	182
Teacher (Kristen):	Studio? All right, let me try to replace that with Studio.	183
	Oh, now we've got blue and pink. What does the blue	184
	represent?	185
Student (unknown):	DreamWorks.	186
Teacher (Kristen):	DreamWorks, how are you making that decision?	187
Student (unknown):	Top right.	188
Teacher (Kristen):	Okay, over here in the legend, we can see the colors.	189
	Okay. Something else I'll point out that is super helpful.	190
	If I take that studio button, and I also put it on the left	191
	here [dragging the "studio" attribute to the y-axis. See	192
	Figure 5.4.], watch what it does for me.	193
Students (chorus):	Oooo.	194
Teacher (Kristen):	Now we probably we can see that there's far more	195
	what?	196
Student (unknown):	DreamWorks.	197

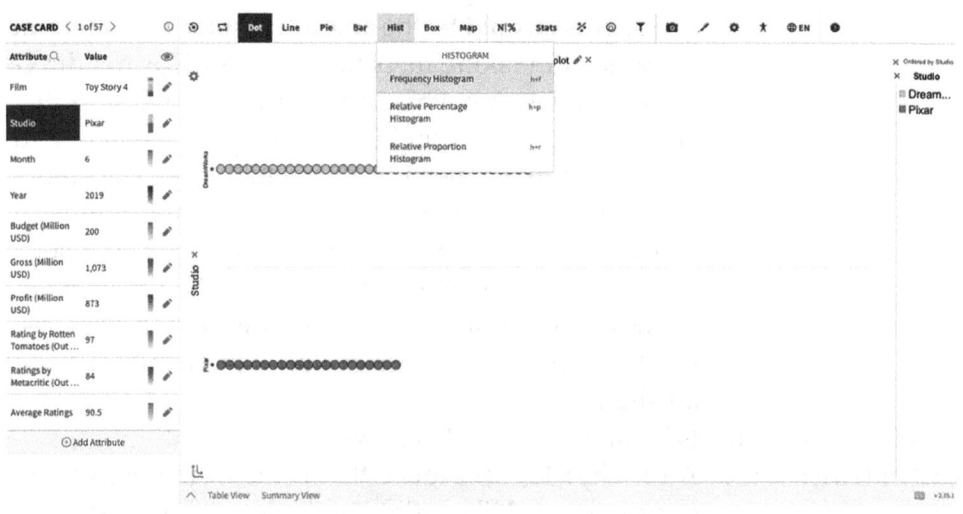

Figure 5.4 Studio attribute dragged and dropped on the y-axis.

Teacher (Kristen):	DreamWorks than Pixar in this dataset. Right? Okay,	198
	the other thing that we might want to think about is	199
	maybe as we're making some arguments here. So again,	200
	we're trying to make a claim based on what?	201
Student (unknown):	Ratings?	202
Teacher (Kristen):	Okay, you might want to do ratings, what was the other	203
	thing we have access to that we can use to compare we	204
	can revenue or gross income or budget, right, those are	205
	all things we can consider. So if I want to use one of	206
	those, what's great about this is let's say I take the gross	207
	amount and drop it down here. [She drags and drops	208
	the gross attribute on the x-axis. See Figure 5.5] All of	209
	a sudden, it plots it for me. For each one of these. Any-	210
	thing you're noticing right away just by looking at it.	211
Students (chorus):	Pixar makes more money.	212
Teacher (Kristen):	Pixar seems to be making more money is that what	213
	you're saying? So down here are some other features	214
	I want to point out. Down here you have a table view.	215
	[Pointing to and clicking on the "table view" tab below	216
	the graph.] This table just has a list of all the data, okay?	217
	Summary view it'll tell you some basic summary, but	218
	right now, we haven't done anything to give it a sum-	219
	mary. Okay? So I'm going to point out some of the	220
	stuff at the top up here. Okay? One of the things is this	221

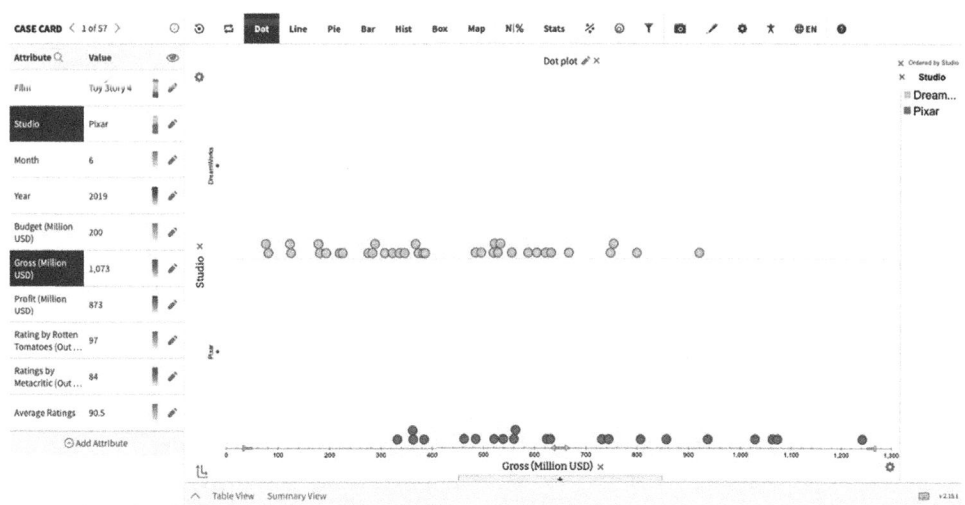

Figure 5.5 Stacked dot plot of gross by studio.

button right here, you see this button? [Pointing to the   222
reset button.] What does it say? Reset the plot. If I click   223
that, look what happens. Now, I would have to go back   224
and do that. But guess what, I'm going to just hit my   225
Undo button, which is right next to it. Okay, so that's   226
reset and undo. So if you try something and you're not   227
happy with the data you're looking at, you're trying   228
to make a different thing happen. You have those two   229
buttons. Then you have these up here. [Pointing to the   230
row of tool buttons along the top.] For instance, what if   231
I want a histogram? Well, I kind of want to do a good   232
reset first. There we go. What if I want to look at this   233
as a histogram instead of just points on a line? Let's   234
choose a frequency histogram for funsies. What do we   235
have? All right. [See Figure 5.6.]   236

Student (unknown):   It's less informative by a lot.   237
Teacher (Kristen):   Do you think it's less informative than the previous one?   238
Student (unknown):   Yeah. Way less.   239
Teacher (Kristen):   What about a box plot? Would a box plot be more?   240
Student (unknown):   No.   241
Teacher (Kristen):   Look at what happens when I hover over the box plot   242
                      though. [As she hovers over the box plot the summary   243
                      statistics are shown. See Figure 5.7.]   244
Student (unknown):   It's more informative.   245

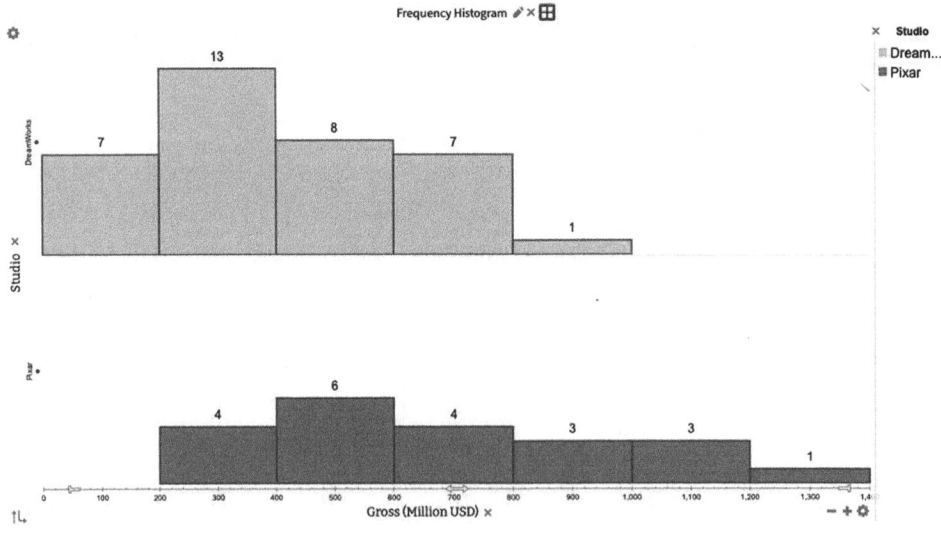

Figure 5.6 Stacked histogram of gross by studio.

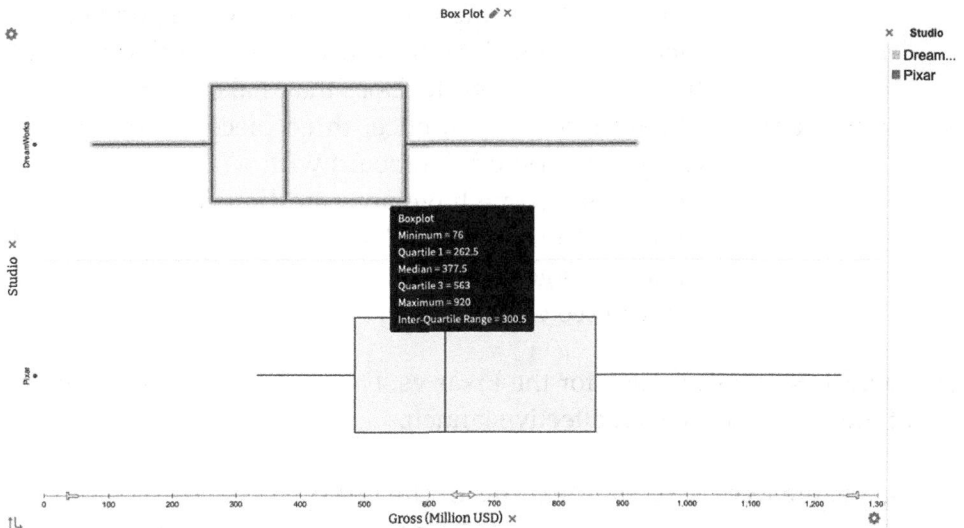

Figure 5.7 Stacked box plot with cursor on plot to show summary statistics.

Teacher (Kristen):	It gives me some new data, right? Okay. The other thing	246
	I want to point out, is you've got this stats button. Let's	247
	say we didn't have a box plot. I click stats, and I click I	248
	don't know what's mean?	249
Student (unknown):	The average or whatever.	250
Teacher (Kristen):	Okay, so we've, we've got this word of average that we	251
	can really use there. And you see it plotted these little	252
	lines for us. You can hover over them and it'll give me	253
	information. Is everybody seeing how we can use the	254
	features here?	255

[At this point Kristen shares a link that goes to the activity sheet where students will record their claims, evidence, and reasons.]

Teacher (Kristen):	So the first thing is, here's my question I'm trying to	256
	answer, which is the best animation studio of these	257
	two, and why? Okay. CER strategy. First thing is the C	258
	stands for making a claim. Okay, so that's what you're	259
	gonna do, you're gonna make a claim, you're gonna	260
	say this was the best and you're gonna tell me why.	261
	Okay? The second part is your evidence. Now, evi-	262
	dence probably looks a little different in here than it	263
	does in English when you use this strategy. Okay? Evi-	264
	dence in English? You're doing the same kind of thing.	265

	But instead of using textual pieces, you're gonna use a	266
	picture of your data or your information that you got	267
	from the Tuva website. Does that make sense?	268
Teacher (Kristen):	Three pieces of evidence, three pieces of reasoning.	269
	Okay. Is everybody on board with what I want you to	270
	do? Does anyone have any questions or need clarifi-	271
	cation on something? Okay, go ahead and get started	272
	I would suggest playing around in the Tuva first so that	273
	you can see what claims you can make	274

Let's unpack Kristen's launch for the Pixar vs. DreamWorks task in terms of each of the four aspects of an effective launch.

Discuss the key contextual features of the task → By taking the time to ask students about their favorite Pixar and DreamWorks movies as well as showing the video compilation, Kristen is helping to make sure the students understand the context of the task.

Discuss the key mathematical ideas of the task → By noting that the students are drawing on their personal connections to the movies rather than data, they are discussing the key mathematical (statistical) ideas of the task. In addition, as her students are identifying the various attributes, tools, and representations related to the Pixar and DreamWorks data in Tuva, they are discussing the various ways they might use the data and tools provided to explore the question that was posed.

Develop common language to describe key features → When Kristen asked the students what data they might want to use to determine which studio was more successful the students noted they would want to know which movies were most popular, had high ratings, and how much money they made. Then when they got into the data, she asked what attributes were in the data set and made the connection between the Rotten Tomato and Metacritic data and their idea of looking at ratings. In addition, she explicitly asked students to consider the names of the attributes and which others might not be familiar to open the floor for discussion about what each of them means in context. There was a back-and-forth between Kristen and the students in which they made connections between the attributes they had previously mentioned and those that were included in the data set.

Maintain the cognitive demand → As the task was launched, the specific question they were to explore was made clear, but no expected strategy was shared. In fact, Kristen was careful to note that people might have different responses based on the data they used to support their claims. So the cognitive demand was maintained.

This particular launch took longer than most launches because it was important to make sure the students were familiar with the context and the data in addition to the fact that the students were being introduced to a new type of technology. The next time the students use Tuva, the launch will be much shorter as she will not need to go in depth when showing students where to find the tools and how to use them. However, because of her careful launch, the students all knew what was expected of them and were comfortable enough with the technology to jump in and get to work. They completed the investigation and shared their findings in a single class period.

Example 2: Arcs and Intersecting Chords

It is common to use preconstructed geometry sketches to explore geometric relationships prior to proving them. In this example, Kristen used a preconstructed GeoGebra sketch with accompanying prompts on a worksheet (Figure 5.8, Link 5.4). The goal was for students to determine the relationship between the measures of arcs and the angles formed by intersecting chords.

Link 5.5 Arcs and Intersecting Circles

> **PAUSE AND CONSIDER**
>
> Look over the Arcs and Intersecting Chords task (prompts and sketch). The sketch is created in GeoGebra. Points C, D, E, and F are all draggable and the measures shown change accordingly. How would you launch this task? Consider all four elements of an effective technology-enhanced task launch.

Kristen invited us into her classroom to observe her teaching this lesson. Let's take a look at how she launched the Arcs and Intersecting Chords task. The transcript of the task launch follows. As you read, identify what you think are key aspects to the success of the launch.

Link 5.6 Video Clip: Launch of the Arcs and Intersecting Chords Task

Circle Angles Exploration   Name: _____

In this investigation, we will take a look at other angles formed inside or outside a circle. We will be using a Geogebra workbook that can be located at: www.geogebra.org/m/FmQ3qgjq

Investigation 1:
Open *Arcs Intersecting Chords* on the Geogebra site above and answer the following questions.

1. What geometric parts of a circle are present? What else do you notice?

2. Move the C, D, E and F and record some information in the table.

$m\overset{\frown}{CE}$	$m\overset{\frown}{DF}$	$m\overset{\frown}{DF} + m\overset{\frown}{CE}$	$m\angle CGE$	$m\angle FGD$

3. How do the values in the highlighted columns of the table relate to each other?

4. Write an equation that represents your conjecture to question 3. Provide a sketch to accompany it.

Given: A circle with two intersecting chords.

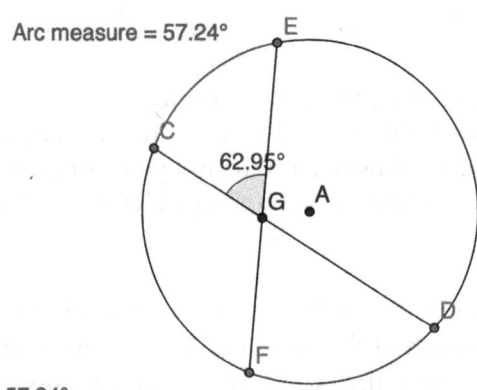

Arc measure = 57.24°

62.95°

Arc measure = 68.67°

Arc CE = 57.24°
Arc DF = 68.67°
Sum of arcs (CE + DF) = 125.91°

Figure 5.8 The arcs and intersecting circles task

Vignette: Kristen Launching the Arcs and Intersecting Chords Task

Teacher (Kristen):	Okay, so this was the image you guys were presented with for arcs intersecting chords. Okay. [displays the sketch full screen] First question just asked, what do we see in the picture? Anybody want to give me what they wrote down what they saw in this image? What were some of the pieces we saw using our new vocabulary?	1 2 3 4 5 6
Student (unknown):	Central angle, center.	7
Teacher (Kristen):	Central angle, center.	8
Student (unknown):	Vertical angle.	9
Teacher (Kristen):	Vertical angles.	10
Student (unknown):	Inscribed angles.	11
Teacher (Kristen):	Inscribed angles.	12
Student (unknown):	Arcs.	13
Teacher (Kristen):	And arcs. Okay, so help me out. Help me label those. So the first thing she told me was the center, where's the center at?	14 15 16
Students (chorus):	A.	17
Teacher (Kristen):	A [pointing at the center point A]. The second thing she told me was vertical angles. Where do we see vertical angles at?	18 19 20
Students (chorus):	CDE and FGC.	21
Teacher (Kristen):	Love it, he gave me three letters C, G, E [tracing the angle with her finger on the display], and F, G, D [tracing the angle], those are vertical angles, right? So we know those [pointing out the vertical angle pairs] have to be?	22 23 24 25 26
Students (chorus):	Equal.	27
Teacher (Kristen):	Equal to each other. Fantastic. Okay. The next thing is only what came after that?	28 29
Student (unknown):	Inscribed angles.	30
Teacher (Kristen):	Inscribed angles, where do we see inscribed angles? (pause) Key?	31 32
Key:	F, G, D?	33
Teacher (Kristen):	F, G, D [tracing the angle with her finger] is that inscribed?	34 35
Students (chorus):	No.	36
Teacher (Kristen):	Define for me inscribed again, Alisa.	37
Alisa:	[can't be heard]	38
Teacher (Kristen):	So G [pointing at G] would have to be where?	39

Alisa:	[can't be heard]	40
Teacher (Kristen):	G, F, D [tracing the angle with her finger, leaving her finger on D]? That could be an inscribed angle. Now, is it really present in the picture? So it might not be important, but I could make an inscribed angle there. Right? Okay. So I could go E, F, D [tracing the angle], right? Or G, F, D [tracing the angle]? I probably use E [pointing at E] because it's on the circle. Right? Okay, but it's not necessarily present. So it might not need a ton of my attention. Right? What was the last thing you told me? I think it was more on your list.	41 42 43 44 45 46 47 48 49 50
Student (unknown):	Arcs.	51
Teacher (Kristen):	An arc? Do we see some arcs?	52
Students (chorus):	Yes.	53
Teacher (Kristen):	What are the arcs? Things seem to be pointing out to us?	54 55
Students (chorus):	[Chorus naming various arcs]	56
Teacher (Kristen):	C, E and D. F. Right? Okay. One thing I want to point out for those of you that haven't tried this, because some of you came across this issue, I think, like Addison and Karina I remember yesterday, right, what do we got to be careful of on this one?	57 58 59 60 61
Student (unknown):	The rounding.	62
Teacher (Kristen):	Say that again.	63
Student (unknown):	The rounding.	64
Teacher (Kristen):	So one thing to pay attention to is there's some rounding happening. So when you're trying to discern what's happening with the values making your pattern prediction, you might want to pay close attention, the rounding might be slightly off. Okay. So don't feel like that last decimal is not perfect. It might be just an issue with rounding. Okay. What was another thing we had to look out for today with somebody else had this issue? I think Kayla and Martin had an issue as well. Maybe I'm wrong. Do you remember? You guys had an issue?	65 66 67 68 69 70 71 72 73 74 75
Martin:	Yeah.	76
Teacher (Kristen):	Do you remember what it was?	77
Martin:	I just know the values came off.	78
Teacher (Kristen):	The values came off. Now be careful when you're rotating these [drags E clockwise around the circle] you can	79 80

move these wherever. But if you start doing things like 81
going all the way around the circle, sometimes [drags 82
C counterclockwise around the circle] … it didn't do it 83
that time, but I promise crazy things will happen. Or if 84
you're making sure they're not crossing. Right. There's 85
some things. Here might be an issue [C is back in its 86
original position]. See, I've made them cross it. No, that 87
didn't do it. I don't know some of you have made it do 88
it. So I don't know how you did it (laughing) because 89
I thought I could recreate it. But just be careful about 90
how you're rotating. Maybe don't keep going around 91
and around the circle. Maybe just slightly alter it so that 92
you don't come up with erroneous measurements here. 93
Okay. Um, all right. That's pretty much it on that. 94

Let's unpack Kristen's launch for this task in terms of each of the four aspects of an effective launch. While Kristen would tell you her launch was not perfect, she did attend to each of the aspects. A breakdown of each aspect is provided next.

Discuss the key contextual features of the task ➜ By asking the students to take a few minutes to simply identify what they see in the sketch Kristen is helping to make sure they understand the context of this task. While this task does not have a "real world" context, it is still important to make sure they understand what it is they are exploring.

Discuss the key mathematical ideas of the task ➜ As Kristen and her students are identifying the contextual features of the sketch, they are naming them using both the labels in the figure (with Kristen pointing each out) and precise mathematical language. Kristen also made sure that she dragged the points on the circle around the circle to show that they are draggable and that the measures change as they are dragged. It is notable that while a student correctly points out that there are inscribed angles in the figure, Kristen notes that since they aren't clearly identified they might not be as important to this exploration. This move validated the student's observation while also helped keep students focused on the important mathematical ideas specific to the investigation at hand.

Develop common language to describe key features ➜ Kristen launched the task by asking students what they saw in the sketch. She used what they pointed out to drive the launch. There was a back-and-forth between Kristen and the students in which they made connections between the specific mathematical objects they had previously defined and how they were represented in the sketch.

Maintain the cognitive demand ➜ As the task was launched there was no specific discussion about what is being investigated. Not even a mention of the "highlighted columns in the table" that will be the focus of their investigation. So the cognitive demand was for the most part maintained. However, there was one move Kristen made that might have an effect on the cognitive demand was asking students what they know about vertical angles as that relationship will likely be important to their investigation. However, her hope was that in bringing that up it was just a reminder to look at relationships they had studied prior.

Looking Across the Examples

Looking across the examples provided in this chapter, it is clear that while there are key characteristics of an effective launch these can be carried out in a variety of ways. Typically they begin with a question that will provide you some insight into students' familiarity with the context (whether the task is "real world" or not) and/or the key mathematical ideas (e.g., What do you notice? What do you wonder? What do you see in this figure? What does "mean" measure? What is your favorite animated movie?). As you think about the context and key mathematical ideas, consider the ways in which the technology interacts with both. In doing so, make sure students know what is dynamic and how, as well as where to find any tools they may need. Importantly, we see that the launch is a discussion. You are not giving instructions; rather, the back-and-forth is necessary as that is how common language is developed. Finally, lowering the cognitive demand is easy to accidentally do! That is why it is so important to think about your launch ahead of time. With careful planning, your students will be launched into the task, and rather than having to run around and address all the hands that immediately pop up into the air, you'll be able to move directly into carefully monitoring your students' as they work.

> **CHAPTER TAKEAWAYS**
> Planning an effective launch is imperative to the success of your technology-enhanced task. As we saw in the examples in this chapter, there is not one way to launch a task but effective launches have similar characteristics.

They discuss the key contextual features of the task, discuss the key mathematical ideas of the task, develop common language to describe key features, and maintain the cognitive demand of the task.

Planning questions to help launch a technology-enhanced task effectively.

Mathematical Goals of the Lesson
- What are the mathematical goals for this lesson?
- What prior math understanding and skills does this build on?
- What is the new mathematics developed by this task?
- How are the dynamic features of the technology intended to support this development?

Key Contextual Features of the Task
- What are the key contextual features of the task? Keep in mind that sometimes the context is something being represented using the technology (e.g., a particular figure or data set).
- Which features are likely to be unfamiliar to some or all of my students?
- How will I elicit and develop students' understanding of these features?

Key Mathematical Ideas of the Task
- What key math ideas do my students need to understand so that they will be able to engage in solving the task?
- How can the key math ideas be represented using the technology? In what ways can students act on those objects, and what will they see in response to those actions?
- Which ideas and representations are likely to be unfamiliar to some or all of my students?
- How will I elicit and develop students' understanding of these ideas?

Development of Common Language
- Which additional language in the task statement (including the technology sketch) is likely to be confusing or unfamiliar to some or all of my students?
- How will I support my students to develop common language to describe the key contextual features, mathematical ideas, and additional language central to the task?

Maintain the Cognitive Demand
- What specifically do I need to avoid doing in the launch so that I maintain the cognitive demand of the task?

(adapted from Jackson et al. (2012), p. 29)

Questions to Discuss With Your Colleagues

1. In what ways might the launch of a task support each and every learner in engaging in the task?
2. Think back to a task that you launched that did not go very well. Based on what you have read here, what do you think you could have done differently? Explain.
3. Think about a technology-enhanced task that you hope to use in a lesson someday. Talk through the planning questions provided earlier. Then talk through how you would actually launch the task.

 You can find links to all the technology-enhanced tasks and supplementary videos throughout the book at https://www.tlmtresearch.com/teachingmathtechbook.

6

Noticing and Eliciting Student Thinking in Technology-Mediated Environments

Noticing and eliciting student thinking are foundational teaching practices (NCTM, 2014) as they inform the decisions we make during instruction. Once we have launched students to begin working on a task, we need to be out and about in the classroom using our eyes and ears to gain insight into their thinking. In order to use student thinking to strengthen instruction, it is necessary to first *notice* student thinking. Noticing student thinking is described as "planning for ways to elicit information, interpreting what the evidence means with respect to student learning, and then deciding how to respond on the basis of students' understanding" (NCTM, 2014, p. 53). This is especially important (and complex) when students are working on a technology-enhanced math task as the ways they interact with the technology (and what they see as a result of that interaction) provide insight into their mathematical thinking.

An Example: Mystery Transformations

Imagine you are the teacher in a secondary classroom where students are working on a "mystery transformations" task. Students have been studying transformations (e.g., translations, rotations, reflections, and dilations) and have been challenged to determine the transformation that was applied to map a pre-image to an image in a dynamic geometry environment.

DOI: 10.4324/9781003302285-7

Link 6.1 Mystery Transformations

PAUSE AND CONSIDER

What would you plan to watch/listen for as you monitored students' working on this task? Explain.

In the following vignette, you will see a pair of students, Addison and Abby, trying to identify the transformation that maps triangle *ABC* to triangle *A'B'C'* (Figure 6.1). Imagine you just walked up behind them and pay close attention to their thinking.

Link 6.2 Vignette Video: Abby and Addison Working on the Mystery
    Transformation Task

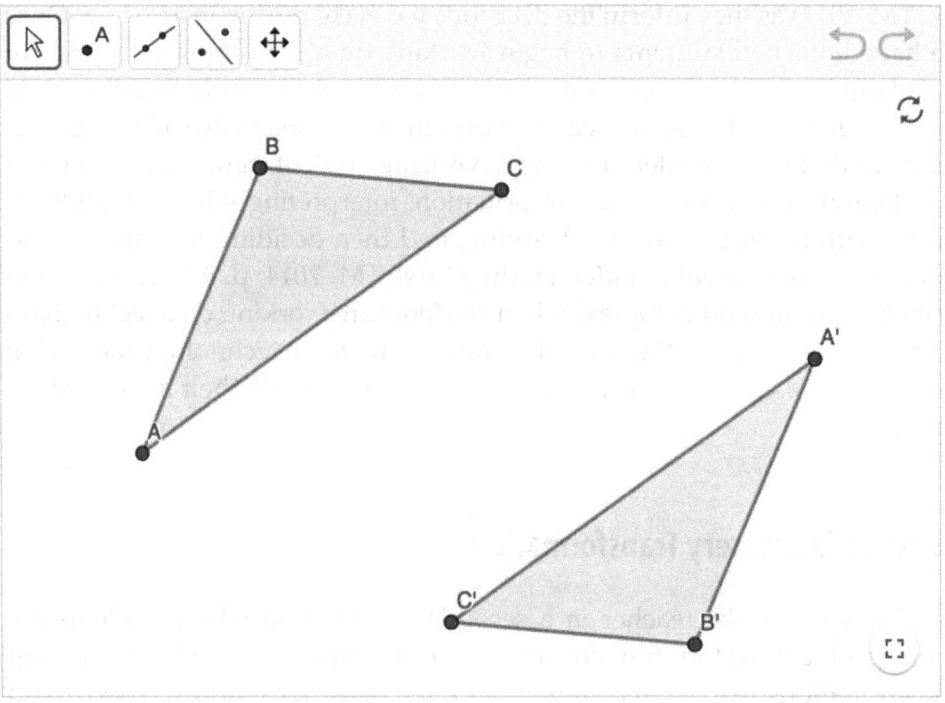

Figure 6.1 Mystery transformations triangles.

Vignette: Abby and Addison Working on the Mystery Transformation Task

Addison:	[Abby constructs line EF in between preimage triangle ABC and	1
	image triangle A'B'C'] No, no, no, no, no.	2
Abby:	Stop, stop. [Addison takes the laptop from Abby. Under the	3
	transform menu, she chooses reflection, chooses line EF and	4
	then triangle ABC. Image $A'_1B'_1C'_1$ is reflected over the line as	5
	shown. See Figure 6.2.]	6
Addison:	It's not gonna be exactly the same.	7
Abby:	But it's gonna be close.	8
Addison:	You're not very good at this.	9
Abby:	I know that.	10
Addison:	That's good.	11
Abby:	Look, see.	12
Addison:	But it's not exactly the same.	13
Abby:	Okay. It's not gonna be.	14
Addison:	I can make it exactly the same.	15

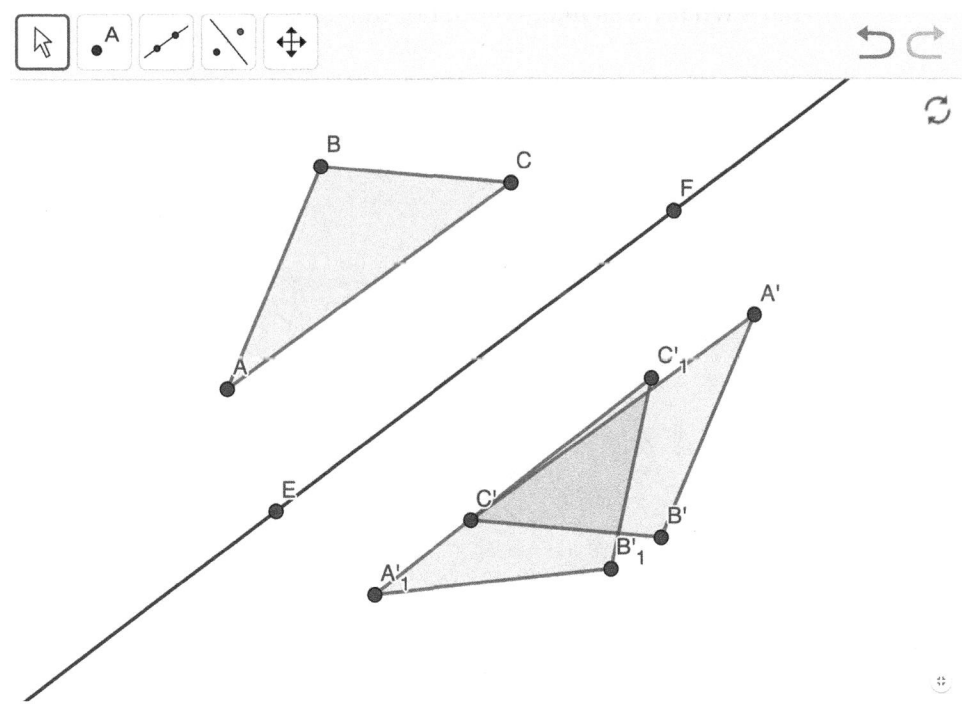

Figure 6.2 Triangle ABC reflected over $\overline{EF}$.

Abby:	Oh, wait. [Abby takes the computer and chooses the select tool.]	16
Addison:	I can make it exactly the same. [Abby drags triangle ABC little	17
	bits trying get the two image triangles to align]	18
Abby:	I don't care. … Look at that. It's very close. … It's extremely	19
	close. [Addison takes the laptop back and clicks undo until the	20
	line of reflection, *EF*, disappears. Addison drags point C onto	21
	point A′, which causes point C′ to lie on point A. She then con-	22
	structs line *AC*.]	23
Addison:	Move, please.	24
Abby:	You're so extra.	25
Addison:	You got something to say?	26
Abby:	Yes.	27
Addison:	[Addison reflects triangle *ABC* over line *AC* and triangle $A_1'B_1'C_1'$	28
	appears as shown in Figure 6.3] Oh!	29
Abby:	Nice. Real smooth. Look at that.	30
Addison:	No, why isn't it the same?	31
Abby:	[Abby takes the laptop back from Addison and undoes the	32
	reflection Addison constructed] Because you did the line wrong.	33
	That's what I was like, you're not that smart.	34

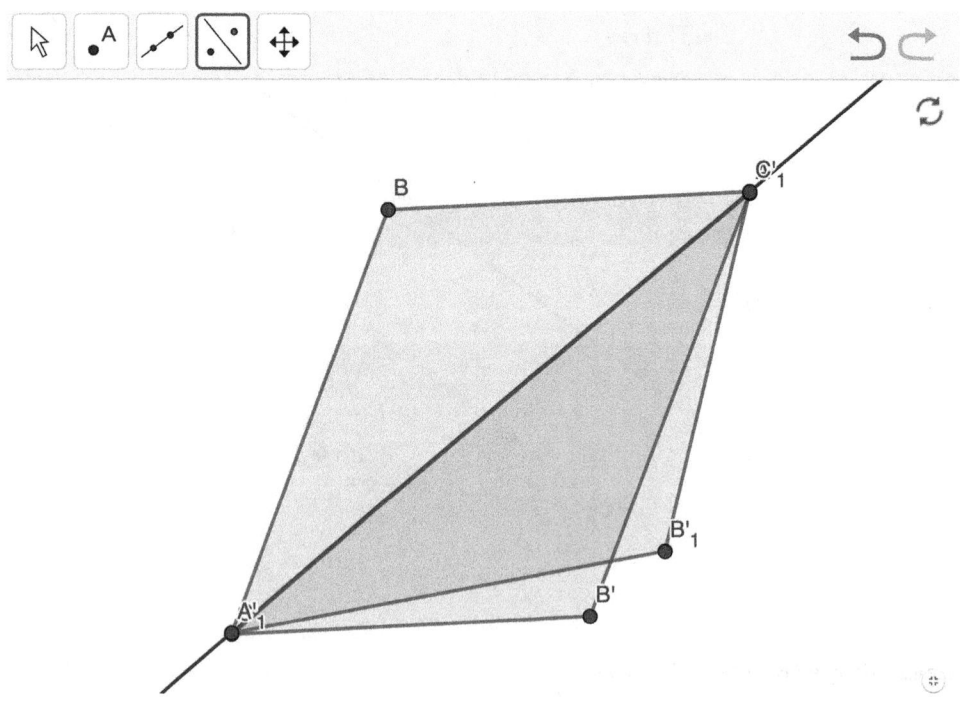

Figure 6.3 Triangle ABC reflected over $\overline{AC}$.

Addison:	I am too!	35
Abby:	The line is not there! That's not where the line is.	36
Addison:	Where is it then?	37
Abby:	[Abby drags point A, trying to locate a line of reflection that works.] All I know is it's not there … because that did not work.	38 39
Addison:	Obviously.	40
Abby:	Wait, oh wait, you made, you made a perfect quadrilateral, right? [After some additional dragging, Abby drags the vertices again so that the two triangles form a quadrilateral.]	41 42 43
Abby:	Now that we've done that … Darn it. I get it, Addison. Stop giving me grief! [Abby reflects triangle ABC over the diagonal of the quadrilateral.]	44 45 46
Addison:	You still didn't do it.	47
Abby:	[thumping her hands on the table in frustration]	48
Addison:	Okay, let's see let's see let's see let's see.	49
Abby:	Okay.	50
Addison:	[Addison takes the computer from Abby and drags point B in several directions to see what happens.] Okay. Let's see why.	51 52
Abby:	Wait, wait, wait! Yeah, move it. Yeah. Oh, what are you doing? … What were you saying Addison? … Wait no no no, move it the other way. Other way.	53 54 55
Addison:	[Addison drags B so that B' and B'$_1$ almost lie on each other. See Figure 6.4.] Hold on I'm trying to make it so it's in screen.	56 57
Addison:	But it's still not an exact copy.	58
Abby:	No, but it's a rhombus, over the line	59

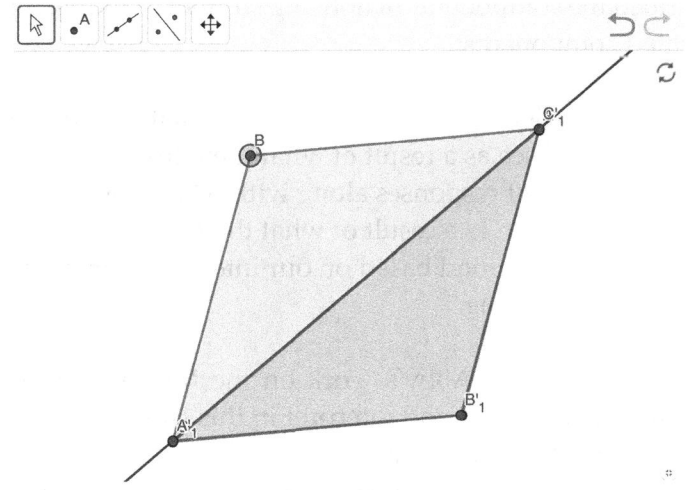

Figure 6.4 Triangle ABC dragged so that B' and B'$_1$ coincide.

PAUSE AND CONSIDER

1. What did you pay attention to as you monitored Addison's and Abby's work?
2. Did you focus on what was said? Explain.
3. Did you focus on how the technology tool was used? Explain.
4. Did you make connections between what Addison and Abby said and what they did/saw with the technology? Explain.
5. Did you pay attention to the affordances and limitations of the technology tool? Explain.

Unpacking Abby's and Addison's Thinking on the Mystery Transformations Task

When students are engaged in a technology-enhanced task and we are attending to their thinking as we monitor their work, too often the focus is on what the students say aloud or write down. When we do this, we are missing an important part of students' thinking – what they are doing and seeing as a result of what they do with the technology. When students are working on a technology-enhanced task, the ways they interact with the technology can provide insight into their mathematical thinking and learning (e.g., Dick et al., 2021; Dick et al., 2022; Dick & Hollebrands, 2011). Noticing student thinking in such a context requires paying attention to not only what students say and do but also the way they engage with the technology. Based on research focused on teachers' professional noticing (e.g., Amador et al., 2020; Jacobs et al., 2010; Thomas et al., 2015) and research on students engaged in technology-enhanced math tasks (Dick et al., 2021; Dick et al., 2022), researchers have described the components of noticing student thinking. This includes three interrelated components:

- *Attending* to student responses along with what they do with technology (and see as a result of what they do)
- *Interpreting* student responses along with what they do with the technology (and see as a result of what they do)
- *Deciding* how to respond based on our interpretation of the student's current understanding

Let's consider Addison and Abby's work on the Mystery Transformations Task. There is a lot to attend to and interpret in this short clip. To help attend to everything that happened in the video we have created a list and separated it by the students' written and spoken work and their engagement with the technology (Figure 6.5).

Attend to Students' Responses	Attend to Students' Technology Engagement
• They say it is not "exactly the same" but say "it isn't going to be exactly the same." • The students say the two images are "pretty close / very close / extremely close." • The students say that the two images do not match up because "you did the line wrong." • The students say "all I know is the line is not there because that did not work obviously." • The students say "you made a perfect quadrilateral." • When the image and preimage still do not line up, they say, "let's see why." • The students say "it's still not an exact copy; no, but it's a rhombus over the line."	• The students construct a line EF between the image and preimage and reflect the preimage over the line EF. • The students drag the preimage to try to line up the sides of the two images. • The students delete the line and drag the points on the preimage to form a quadrilateral in which C and A' and A and C' intersect. • The students construct a line that is the diagonal of the quadrilateral and goes through the shared base segment and then reflect the preimage over the line. • They continue to drag A bit by bit. • The students delete the reflection and drag the points of the preimage to form a quadrilateral. • The students construct a line that is the diagonal of the quadrilateral and then reflect the preimage over the line again. • The students drag point B to determine why the reflection did not work. • They end with B so that they have almost created a rhombus.

Figure 6.5 Attend to students' responses and their technology engagement.

If we only attended to what Addison and Abby said (left column), we could interpret that they understand that a reflection should produce a congruent triangle and that if the line of reflection were the side of the triangle it would be the diagonal of a quadrilateral formed by the preimage and image. But from watching the video/reading the transcript we know that there is so much more we can say about what they do and do not yet understand. If we coordinate our attention to their spoken responses (left column) and how they engaged with the technology (right column), we can construct a more complete picture of their understanding. Based on Addison's and Abby's engagement with the technology, we see that they do not appear to understand if the transformation was a reflection of the pre-image and the image should not only be congruent triangles, but the image A'B'C' and their constructed image $A_1' B_1' C_1'$ need to also have the same orientation. There is no evidence that the students ever considered the orientation of the two images as the vertices they are trying to align are labeled differently (e.g., A' and C_1') and they do not mention that difference.

In addition, Addison and Abby do seem to have an understanding that the preimage and the image should be equidistant from the line of reflection. This was based on their placement of the line of reflection between the two triangle bases. However, since they did not consider point B it does not seem that they understand that all vertices of the triangle and their images must be equidistant from the line of reflection. We have a much deeper understanding of Addison's and Abby's understandings when we coordinate what they said with how they engaged with the technology.

Now that we have a more complete description of their understanding we can decide how to respond. One decision we could make is to probe their understanding related to the pre-image and images being equidistant from the line of reflection by asking them to use the GeoGebra tools to determine if the vertices of the triangles are in fact equidistant from the line of reflection. How they measure the distance of B, B′, and B_1′ from the line of reflection will reveal a lot about their understanding of this property of transformations. But not all of our decisions of how to respond have to include technology. We could also ask Addison and Abby to draw a mapping diagram from the vertices of the pre-image to the given image and then the pre-image to their constructed image, which could help to draw their attention to the orientation of the three figures.

A Framework to Guide Our Noticing of Student Thinking

Given that noticing student thinking on technology-enhanced tasks is a complex practice we have included a framework to help guide this work. The Noticing Students' Mathematical Thinking in Technology-Mediated Learning Environment (NITE) framework (Dick et al., 2021) was designed to demonstrate the coordination needed between student responses and how they engage with the technology and to support teachers' noticing by providing an easy image to keep in mind while working on this important practice (Figure 6.6).

The arrows in the NITE framework indicate that all components of noticing are by their nature interrelated (Jacobs et al., 2010). However, *attention to* and *interpretation of students' responses* (what they say and/or write) are separated from *attention to and interpretation of students' technology engagement* (what they do with the technology and what they see as a result of what they do) to highlight the importance of coordinating the two. The *decide how to respond* component is separated from the other components to balance the importance of focusing on both student responses and their technology engagement prior to making instructional decisions. In addition, when deciding how to respond one must consider how to position the technology (or not) in their response to support students in moving their thinking forward.

Drew, a Tech-Math Teacher who has used the NITE framework, shared that he keeps the framework up on his computer monitor when he watches

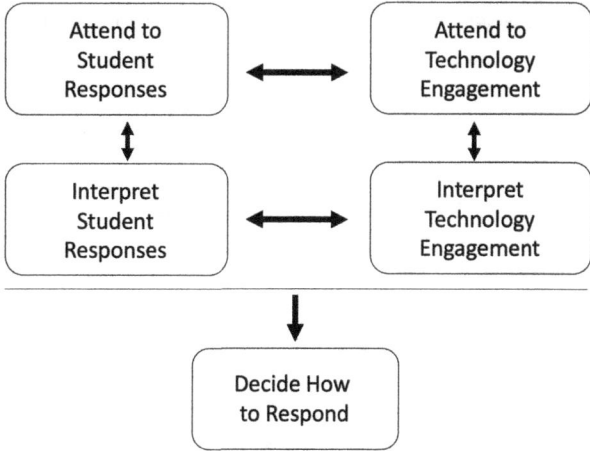

Figure 6.6 The NITE framework.

videos of students working in technology-mediated environments and as he plans his lessons as a reminder to "take in all aspects to truly get a better understanding of what the students know". He shared that when he first started working on noticing student thinking on technology-enhanced tasks, he was

> very focused on what the students know, what are they telling me. I was so focused in on am I getting what they know, but I kind of needed to like step back and look. It's like here is a completely different area that they are giving you information [referring to students' technology engagement] that you're not even considering just because you are trying to zone in on what they are saying. You're trying to hear what they are saying and what they're writing and you're not looking at how they are actually interacting with it.

Keeping the NITE framework in mind, let's look at a few more examples.

An Example: Making Sense of Box Plots

Developing a conceptual understanding of box plots is a topic that is developed throughout middle school, high school, and college courses. The Making Sense of Box Plots activity, built in Desmos Activity Builder, has four box plot matching challenges. In each, students are provided a set of data observations that are dynamic, a box plot that is connected to those data that will change as the individual observations are dragged, and a static box

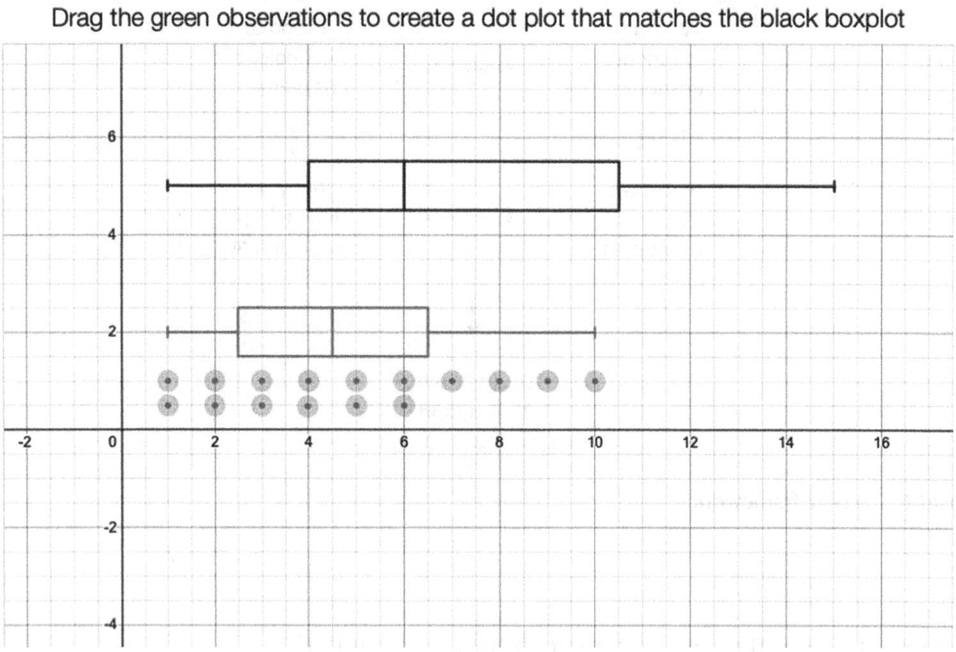

Drag the green observations to create a dot plot that matches the black boxplot

Figure 6.7 Making sense of box plots.

plot (see Figure 6.7). The goal is to drag individual observations to make the dynamic box plot match the static one. This provides an opportunity for students to explore the relationships between data distributions and their different representations.

Link 6.3 Making Sense of Box Plots

PAUSE AND CONSIDER

In the following vignette, a pair of students, JP and Kyle, Integrated Math 1 students who have prior knowledge of measures of center, a five-number summary, and various representations of univariate data (e.g., dot plots, box plots, histograms), are trying to match the dynamic box plot to the static box plot. While you read/watch, keep the NITE framework close by and consider the following:

1. Attend to how JP and Kyle matched their box plot.
2. Interpret JP's and Kyle's current understanding of box plots.
3. If you were JP's and Kyle's teacher and standing beside them, watching and listening, at the moment this video occurred, what would you do next?

Link 6.4 Video: Kyle and JP Working on the Making Sense of Box Plots Task

Vignette: Kyle and JP Working on the Making Sense of Box Plots Task

Kyle:	All right. [JP drags the farthest right point to the location of the max-imum, which causes an outlier to appear.]	1 2
JP:	So, we can keep – so we're gonna have to move this one all the way right here for the maximum. [JP drags a second point to the left of the maximum, causing a second outlier to appear.]	3 4 5
JP:	But then we're also gonna have to move this one right there. [JP drags a third point to the left of the maximum, causing a third out-lier to appear.]	6 7 8
JP:	We're gonna have to move this one too.	9
Kyle:	Mhm.	10
JP:	So if we put three right here, right?	11
Kyle:	I think so. [JP drags a fourth point to the left of the maximum, caus-ing the three outliers to disappear and the fourth-quartile whisker to appear.]	12 13 14
JP:	And then we have to move his one all the way down there.	15
JP:	Yeah.	16
Kyle:	Yeah. [JP moves one point in the third quartile to the right.]	17
JP:	And then we have to move one of these right here.	18
Kyle:	Wait, let's count how many there are total first.	19
JP:	1, 2, 3, 4, 5, 6, 7, 8, 9, 10, 11, 12, 13, 14, 15, 16. Yeah.	20
Kyle:	Okay, so it's still 16 except now this time, the middle is six.	21
JP:	Yeah. [Kyle moves a point in the third quartile left and right.]	22
Kyle:	For this. So I'm guessing we still need to move these down?	23
JP:	No, no, you keep that one there. [Kyle drags a point in the second quartile closer to the median.]	24 25
JP:	'Cause that controls the third quartile. You move that one, but then you're gonna have to move one of these under.	26 27
Kyle:	Huh? [JP moves the point Kyle moved back to the left, and the median moves to the left.]	28 29
JP:	Like this, look. So you see how this one moves that?	30
Kyle:	Mhm. [JP drags a point in the third quartile, and nothing moves on the corresponding box plot.]	31 32
JP:	I see. Wait, I see now. [JP drags a point in the second quartile, and nothing moves on the corresponding box plot.]	33 34
JP:	We're gonna have to move one of these. [JP drags a different point in the second quartile toward the median, and the median moves to the right.]	35 36 37

JP:	And then we can drag it more. [JP drags a point in the second quar-tile, and nothing moves on the corresponding box plot.]	38 39
JP:	And if we move this one ... [JP drags a point in the second quartile toward the median, and the median moves to the right.]	40 41
Kyle:	Since it's not even we're probably gonna need two in the middle again.	42 43
JP:	Yeah. [JP drags a point in the second quartile toward the median, and the median moves to the value of six.]	44 45
JP:	Buh-bam. Right there.	46
Kyle:	Okay, so you could basically say these two are ...	47
Kyle:	In the middle?	48
JP:	Yeah.	49
Kyle:	Yeah. Okay. [Kyle drags a point in the third quartile, and the third quartile moves to the right on the corresponding box plot.]	50 51
JP:	And then to make the, no, you keep that one there though.	52
JP:	'Cause that one's the third quartile and messes it up.	53
JP:	So we have to move these right here. [JP drags a point in the second quartile, and nothing moves on the corresponding box plot.]	54 55
Kyle:	Okay, now.	56
JP:	You're gonna have to move the bottom one too.	57
Kyle:	Yeah. [Kyle drags a point in the second quartile, and nothing moves on the corresponding box plot.]	58 59
JP:	Or actually, not the bottom one. [Kyle continues to drag the point, causing the second quartile to increase in value.]	60 61
JP:	Like the one – Oh, yeah.	62
Kyle:	Okay.	63
JP:	Right there.	64
Kyle:	Okay, and these stay the same.	65
Kyle:	Okay, let's see how many. One, two, three, four. One, two, three, four. One, two, three. One, two, three.	66 67
JP:	Yeah.	68
Kyle:	One, two.	69
Kyle:	Okay, so now we know basically every time we do it with the even number, there should be two in the middle and then the same on each side.	70 71 72
JP:	Yeah.	73
Kyle:	Of the whiskers	74
Kyle:	And the boxes.	75
JP:	Yeah, 'cause it's four on the outside, four on that side and then two in the middle, middle-ish.	76 77

Kyle:	Yeah, I guess if it were an odd number, we would have to –	78
JP:	Do the opposite.	79
Kyle:	Yeah, and there would only be one in the middle.	80
JP:	Yeah.	81
Kyle:	Instead of two.	82

Unlike Addison and Abby, Kyle and JP said a lot for a teacher to attend to their understanding of box plots. However, just like Addison and Abby, the way in which Kyle and JP engaged with the technology also highlighted additional understandings. From listening to Kyle and JP's exchange, we can tell that they are paying attention to the number of data observations and that they know that if there is an even number of observations, they will need to use two to determine the location of the median. However, they seem to believe the two observations points must lie on the median. We also get a sense that they have noticed that particular observations seem to control the location of the quartiles.

It is through careful coordination of what they said and how they engaged with the technology one could interpret that Kyle and JP understand that there needs to be a data observation at the maximum and the minimum as well as other observations nearby to create the whiskers of the box plot. This is evident from both what they said, "We need to move this one all the way right here for the max" and the fact that they continue to drag points toward the maximum value when they see moving only one results in an outlier and not a whisker. In addition, while we hear them count the data observations, without seeing where the observations that are being counted were located, we miss that they touch one of the points they placed at the median without saying "four" as they counted the number of observations in the second and third quartile; without noticing this, we would not have evidence that they understand that each quartile should have the same number of data observations. In addition, we would not have evidence that they have not yet considered how the distribution of those observations relates to the length of each quartile.

After coordinating our interpretations of Kyle's and JP's thinking aloud and technology engagement, we can make an informed decision about how to respond. For example, to help them understand that for an even-numbered data set that two middle observations do not have to lie on the median, one might ask JP and Kyle if they could move the observations to new locations and still keep the median value of six. Similarly, since they placed data observations in the "middle-ish" of Q1 and Q2, one might ask them why they did not have any points near Q3. Finally, one might decide to do nothing at this time. There are two more challenges in the activity, and each is more complex than the prior. Working on the more complex challenge alone might result in revisions to their current thinking about the distribution of data observations in a box plot.

An Example: Rational Functions and Vertical Asymptotes

To introduce vertical asymptotes in the context of rational functions this task, created in Desmos Activity Builder, uses sliders for students to explore the effect of changing the parameters of a rational function in the form $f(x) = \dfrac{k}{ax+b}$, on the location (and existence) of vertical asymptote(s). The first few pages of the activity show various graphs of rational functions with constant numerators and linear denominators with prompts asking students to describe the domain and range and asking what they notice comparing across the graphs. Then the term *vertical asymptote* is introduced and students investigate the relationship between the structure of the function and the existence and location of any vertical asymptotes using sliders with the vertical asymptote shown in the graph along with the function. As sliders are changed, the graph of the function and the function in the directions at the top dynamically change accordingly, as does the vertical asymptote (see Figure 6.8).

Link 6.5 Task: Introduction to Vertical Asymptotes

Use the sliders to change the parameter values in the function. f(x) = -7 / (-1x + 1) How can you predict the location of a vertical asymptote given the function rule?

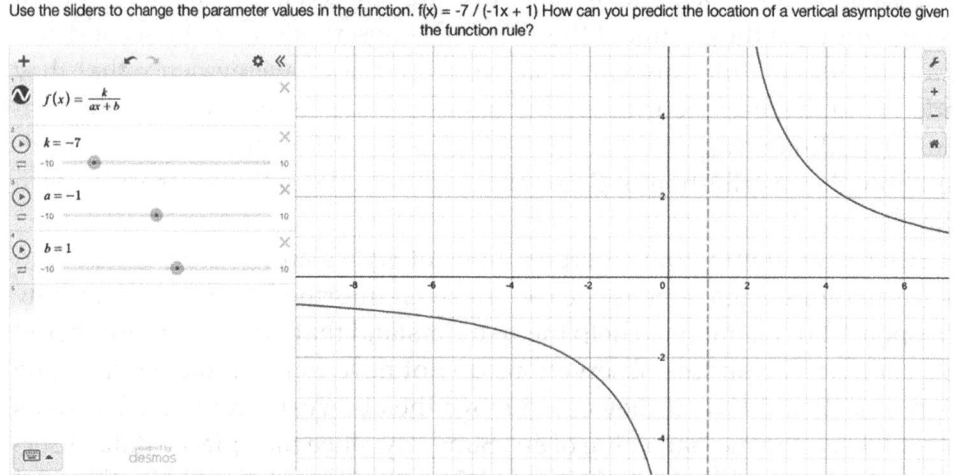

Figure 6.8 Introduction to vertical asymptotes.

PAUSE AND CONSIDER

Keep the NITE framework close by as you read/watch the following vignette. Eden and McKenzie are working on screen 5 of the Introduction to Asymptotes activity. They are using sliders to explore the parameters, k, a, and b in the rational function, and their effect on the location (and existence of) vertical asymptotes. While you read/watch, consider the following questions:

1. Attend to how Eden and McKenzie determined the location of the vertical asymptote for a rational function of the form $f(x) = \dfrac{k}{ax+b}$.
2. Interpret Eden's and McKenzie's current understanding of vertical asymptotes.
3. If you were Eden and McKenzie's teacher and standing beside them at the moment this video occurred, what would you do next?

Link 6.6 Video: Eden and McKenzie Working on the Vertical Asymptotes Task

Vignette: Eden and McKenzie Working on the Introduction to Vertical Asymptotes Task

McKenzie:	The asymptote isn't based on … just those numbers alone. Because look, if you do it there … it's not going to be exactly on that number. [Student drags slider b to change the value from 5 to 4 which moves the graph to the right on the x axis.]	1 2 3 4
Eden:	Yeah …	5
McKenzie:	So … how can you predict the location of a vertical asymptote given the function rule? I think … wait … I don't know … I'm confused because "a" and "b" do the same thing … wait move "a" [Student drags slider a to change the value from 5 to 2, which results the graph moving to the left 3 places and the graph stretches horizontally.]	6 7 8 9 10 11
McKenzie:	I feel like it makes it bigger, or is that just me?	12
Eden:	Oh oops! What did I just do?	13
McKenzie:	Go to zero. [Student drags slider to change the value to 0, which results in a horizontal line at $y = 1$.]	14 15
Eden:	But it's still three.	16
McKenzie:	That moves up and down because … this doesn't move vertical asymptote. [Student drags slider k to change the value from 1 to 12, which does not affect the vertical asymptote but results in the graph stretching vertically.]	17 18 19 20

McKenzie:	This moves your vertical asymptote, so something with that and this one is your vertical asymptote. [Student drags slider a to test values from −1 to 4 which results in the horizontal asymptote moving from left to right.]	21 22 23 24
McKenzie:	So I think that whatever "b" is your vertical asymptote.	25
Eden:	I think it's …	26
McKenzie:	But it has something to do with "a" too, though.	27
Eden:	I think it's … I think it's … um … [groaning] divided by "a". Yeah, I think it's, I think it's "b" divided by "a" 'cause, 'cause look two divided by four is what?	28 29 30
McKenzie:	Two divided by four is one half, but two divided by negative four is negative one half.	31 32
Eden:	I know, I think it's just, I think it's just one of those weird flippy thingy with that graphs do.	33 34
McKenzie:	Let's try this number and … oh and I need to go down to this number. Hold on one … point five. [Student drags the sliders and sets the values at $k = 0$, $a = -10$, $b = 5$. K = 0 results in the graph not moving even when the student changes the a value from −10 to 0.]	35 36 37 38 39
McKenzie:	Oh! You're right 'cause that's point five.	40
Eden:	I think it's one of those weird flippy thingy. That doesn't really make sense and yeah	41 42
McKenzie:	Or … This [laughing] or …	43
Eden:	Two … two and a four	44
McKenzie:	We can do two in here and four. [Student drags sliders a and b and sets the value of a to 2 and the value of b to 8 which changes the vertical asymptote from $x = -.5$ to $x = -4$.]	45 46 47
Eden:	Yeah.	48
McKenzie:	Yeah, it's one of those flip things. And then that's …	49
Eden:	So … if I were to define the vert … I would tell my friend to … divide … "b", whatever "b" is, by "a" and then make it equal … okay, "b" divided by "a".	50 51 52
McKenzie:	One sec, I have to write this out.	53
Eden:	Equals negative "x" or whatever.	54
McKenzie:	Okay.	55

From just the written exchange, it is apparent that they have determined a way to find the location of a vertical asymptote for rational functions of the form $f(x) = \dfrac{k}{ax+b}$, but if a teacher only paid attention to the students' written response, the teacher would not have a full picture of the students' current understanding. Most important, the students set $k = 0$ and the graph of the function "disappears"; they are no longer looking at a rational function while they are dynamically changing the values of parameters a and b and reasoning about the vertical line they see. It is not clear if they don't notice that the function is now $f(x) = 0$ or if they did it intentionally so that they could focus solely on the vertical line that was representing the vertical asymptote.

When we consider what McKenzie and Eden said as well as how they engaged with the technology, we can interpret that Eden and McKenzie do not show evidence of understanding that the vertical asymptote represents where a function is undefined; rather, they are focused on looking for a connection between a, b, and the location of the vertical asymptote. They understand that parameters a and b play a role in their locations; however, they have not yet made a connection between the parameters a and b and setting the denominator of the function equal to zero to find what values of x make the function undefined.

After coordinating our interpretations of McKenzie's and Eden's thinking aloud and technology engagement, we can make an informed decision about how to respond. For example, noticing the students' rule for finding the location of the asymptote was determined when the function they were examining was not rational, you may start by drawing their attention to the function structure and asking them to predict (and then verify using the dynamic graphing tool) what the graph of the function will look like for varying values of k, a, and b in addition to the location of the asymptote. Another possible response would be to ask the students to explain why it makes sense that the asymptote is located at $x = \dfrac{-b}{a}$ given the structure of the function. The first response requires the use of the technology, and the second does not. However, either response would push the students toward connecting the existence and location of an asymptote to the structure of the function equation.

If we were to look just at the students' final responses in each of the examples, we would be missing important insights into their understanding. Through our unpacking of the students' work on these tasks our aim was to make clear why it is necessary to pay close attention to the actions students take with technology and see as a result of these actions when attending to and interpreting students' mathematical thinking. In each pair of students' work we were able to not only understand more about their understanding of the mathematics/statistics but also build off that understanding in how we decide to respond.

The NITE framework can help us think about what we are noticing and coordinating when we monitor students as they are working on technology-enhanced tasks. In addition, it is a helpful tool to keep in mind when anticipating student thinking – pointing us to anticipate the ways in which students will engage with the technology as they are working and how that engagement might inform their thinking.

CHAPTER TAKEAWAYS

- When monitoring students as they are working on technology-enhanced tasks it is important to attend to their written and spoken responses along with what they do with technology – and see as a result of what they do.
- To get a complete understanding of the students' thinking, we need to interpret their responses along with what they do with the technology – and see as a result of what they do.
- When we decide to respond to students, the response should be based on the students' coordinated understanding. We must consider how to position the technology (or not) in our response to support students in moving their thinking forward. Deciding how to respond may or may not include technology use.

Questions to Discuss With Your Colleagues

1. Reflect on the earlier vignettes. What aspects to the students' thinking might you not have picked up on if you had read/watched them prior to learning about the NITE framework? Explain.
2. Think about a technology-enhanced task that you hope to use in a lesson someday. Talk about what components/sketches/pages of the task will elicit opportunities to learn about students' understanding from their engagement with the technology.

 You can find links to all the technology-enhanced tasks and supplementary videos throughout the book at https://www.tlmtresearch.com/teachingmathtechbook.

7

Facilitating Whole-Class Discussions in Technology-Mediated Environments

Throughout this book we have talked about the importance of positioning students as powerful math explorers. Specifically, we have argued that when using a technology-enhanced task, being a math explorer means having ways to act on mathematical objects and observing how they change as a result of one's actions and the importance of student discourse during exploration. Whole-class math discussions around specific tasks provide students an opportunity to develop their mathematical thinking by communicating their ideas, considering the thinking of others, and making connections among the ideas and their prior math understandings. In addition, such discussions position students as valuable resources that promote an equitable learning environment (Smith et al., 2020).

An Example: Discussing Characteristics of Ellipses

Kristen has planned a lesson arc on ellipses for her precalculus class. We are calling this a lesson arc because she designed the activity to play out over at least two full class sessions, possibly longer. To launch the unit on conic sections, she spent a day in the gym having her students do some "challenges" in which they were put into groups of 10 and asked to arrange themselves into shapes with particular characteristics. The challenges included the following:

DOI: 10.4324/9781003302285-8

- There was a cone placed on the floor. The challenge was to arrange the group into a shape where each person was equidistant from the cone.
- There was a tape line and a cone placed on the floor. The challenge was to arrange the group into a shape in which each person is equidistant from the line on the floor I marked and the cone.
- There were two cones placed on the floor. The challenge was to arrange the group into a shape where the sum of each person's distance from themselves to each cone is the same.

After these challenges, the students also looked at conic sections from the perspective of "slicing" a cone, using a conic flashlight. Ms. Fye used a conic light and a poster board for this demonstration (see Link 7.1). Now students will spend the next few lessons exploring each of the conic sections included in precalculus: circles (which they have already studied in Math 3 but are referred to often here), ellipses, parabolas, and hyperbolas. The focus in this lesson is on ellipses.

Link 7.1 Light Cone Simulation

Kristen selected a Desmos activity she found online and adapted to meet her goals. Notice that she has used both a Desmos graphing calculator sketch and a GeoGebra sketch in the activity (Figure 7.1, Screens 3, 5, and 11 all link out to GeoGebra). The link to the activity is provided in Link 7.2.

Link 7.2 Ellipse (Geometric Definitions, Equations, and Graphs) Activity

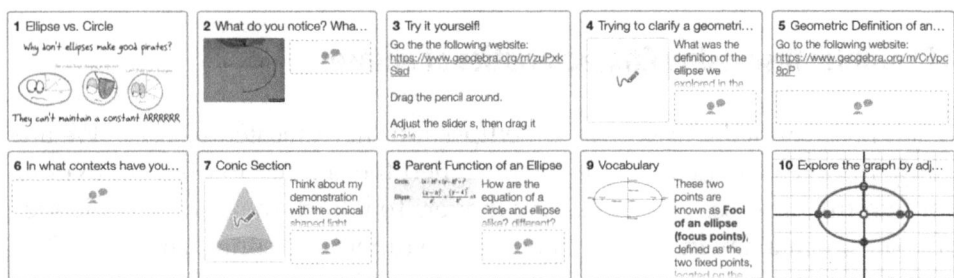

Figure 7.1 GeoGebra sketches linked within a Desmos activity.

PAUSE AND CONSIDER

Take a look through Kristen's Ellipse Activity (Link 7.2).

- At what point(s) in this activity do you think it would be important to stop and hold a whole-class discussion? Explain.
- Imagine this was your class. How might you go about facilitating a whole class with a goal of identifying important relationships between the equation of an ellipse and key characteristics of its graph? Explain.

When using technology-enhanced math tasks, the ways in which we can select and display student work to anchor our discussions are quite different from when students are working with paper and pencil. This is because in technology-enhanced tasks students' strategies are often related to dynamic representations. In this chapter, we move from thinking about how to put structures in place to support student discourse when working in small groups (Chapter 5) to thinking about how to use what we notice when monitoring students' work (Chapter 6) toward facilitating productive whole-class discussions when using technology-enhanced math tasks. We will talk about setting up norms that support collaboration and discourse in your classroom and strategies for using student work to center whole-class discussions.

Norms That Support Collaboration and Discourse

Engaging students as explorers of mathematics includes collaborating with others – learning from others and sharing with others. Yet, the notion of 1:1 devices is often seen as isolating students from each other with their devices. To ensure this doesn't happen, it is important to have routines that support collaboration on technology-enhanced tasks. One of the easiest ways around this is rather than working 1:1, work 2:1 – two students using a single device (if the screen is large, you can even go up to three students on one device). Sharing a screen means that students are looking at the same thing and talking to each other about the same mathematical objects as one of them acts on the object (hopefully taking suggestions from the others or taking turns acting on the object). It is very difficult to *not* talk about math when observing how a mathematical object reacts to the actions you take!

- "Something I love is having two students on one screen. That way it's not isolating the students with the technology, but it's allowing them to have conversations. I know I'll forget to do that in one class, and it'll be like dead silence. Then I'll remember for the next class, and there's just all these conversations going on. So it just makes such a huge difference to get kids talking". (Nolan)

- "So like for me a lot of times I'm hoping kids are talking to each other next to each other even if we're working individually, but let's say I'm very specific about this is a partner activity and you've got the technology in front of you my goal would be to change it to a two to one situation where there's two students on one device because that then forces them to work together. It takes away that element of they can't work individually which forces them to kind of talk and negotiate through the questions that they're going to answer and come to some kind of consensus". (Kristen)

If you have multiple students working with a single device it is helpful to have routines in place to guide how they work together. Most important is coming up with routines for taking turns using the device directly and for making sure that each student's ideas are being explored when using the technology.

- "Maybe we alternate who's inputting the responses so it doesn't become one person dominating. Or we make sure it's part of a routine of where's the Chromebook going to be. It's not going to be over here where someone can't see the screen at all. So there's different routines like that. Expectations that you know from the beginning of the year that you're just you're laying out and you're teaching them how to do". (Nolan)
- "If it's something where I have them recording answers on paper and they're using technology with it I kind of use this driver–navigator scenario with mine. The person with the paper is navigating right now, but they're gonna need a brake. The driver's gonna need a brake later and you'll need to switch. That kind of forces both students to take the onus of both pieces at some point. They may not be doing it the whole time, but that kind of switch helps them a lot". (Kristen)

When switching from small-group discussions to whole-class discussions, it is important to make sure that students are fully engaged and listening to each other. That means we need to have their eyes and ears attending to what is being shared with the whole class, not their own devices. If you are using handheld calculators or iPads, then maybe the routine is that the device is facedown when attention should be elsewhere. If you have laptops, then maybe they are closed or almost closed (e.g., screens at 45 degrees). All of these choices make it easy for you to see your students' faces, for them to look at and listen to others, and for students to easily access the technology again when you are ready for them to do so. Some teacher dashboards have a "pause" feature allowing you to pause students' screens to get their attention (e.g., Desmos Classroom, GeoGebra Classroom).

- "All our students have Chromebook, so it's part of our routine. You know. Close your lids right now, your time for your lids to be closed or not, so there are certain expectations like that and routines that you work through. You're consistent with them". (Nolan)
- "So I would say, put your computer at 45, so that you know when to make eye contact with other people. You know to re-engage in the classroom discussion. Or I just pause the screen. Knowing that when I'm pausing, the screen means that we're going to engage in a dialogue, and that you need to bring your thoughts forward". (Lauren)

With strategies for supporting collaboration and discourse in your classroom in mind, we move to one of the most complex and important practices in a student-centered math classroom – facilitating a whole class discussion.

Facilitating Whole-Class Discussions in the Context of a Technology-Enhanced Task

Throughout this book we have been alluding to a task implementation model organized in three phases – launch, explore, discuss and summarize (Smith & Stein, 2011). This model starts with the selection of a high-level task that is aligned with your learning goals (see Chapter 4). When using the task with students the launch phase is when the teacher introduces the task to the students in such a way that they are ready to meaningfully engage (as discussed in Chapter 5). In the explore phase, students work in pairs or small groups to solve the task, using strategies that make sense to them. They are encouraged to talk about their ideas, including justifications for them, with their group and be prepared to share them with the whole class. Finally, the discuss and summarize phase is when the teacher facilitates a whole-class discussion. During this phase, the teacher uses various student strategies, displayed for the whole class to view and discuss, that highlight mathematical ideas related to their specific learning goals.

The discuss and summarize phase is a challenging one to carry out as effective discussions don't just happen; they have to be carefully orchestrated. After years of observing math teachers holding effective whole-class discussions, Smith and Stein (2011) described a framework to support teachers in doing this work that is referred to as the Five Practices for Successfully Orchestrating Mathematics Discussions (or the 5 Practices for short). This is a slight misnomer as there are actually six practices! The first two (practices 0 and 1) take place when you are planning your lesson and the other four during instruction. They are described in Table 7.1, including what each practice means in the context of using a technology-enhanced math task.

Table 7.1 The 5 Practices With Technology-Enhanced Math Tasks

Practice	Description	What This Means in the Context of Using a Technology-Enhanced Task
Practice 0: Setting goals and selecting tasks	Specifying learning goals and choosing a high-level task that aligns with your learning goals.	If you are choosing to use technology as an integral part of the task, this includes selecting a technology that best supports the learning goals and the representations you want students to have access to create and explore (Chapters 3 & 4).
Practice 1: Anticipating student responses	Exploring how you expect students to solve the task and preparing questions to ask them about their thinking.	This includes anticipating the ways you expect students to engage with the technology, where their eyes might be drawn, what representations they might favor, and how they will make sense of what they see as a result of their actions with the technology (Chapter 4).
Practice 2: Monitoring student work	Looking closely as students work on the task and asking questions to assess their understanding and move their thinking forward.	Monitoring students during the *explore* phase means noticing their thinking using the technology in a coordinated way – attending and interpreting both what they say/write with what they do and see as a result of what they do with the technology (Chapter 6). In addition, some technology platforms allow teachers to monitor all students' work in one place.

Table 7.1 (Continued)

Practice	Description	What This Means in the Context of Using a Technology-Enhanced Task
Practice 3: Selecting student solutions	Choosing solutions for students to share that highlight key mathematical ideas that will help you achieve lesson goals.	When selecting students' technological work you need to think about how to share it publicly. Some technology platforms have tools for selecting specific student work to be displayed for discussion. If the technology used in your lesson does not have this capability you will need to have a way for students to quickly display their work by using a shared document to display images or links.
Practice 4: Sequencing student solutions	Determining the order in which to share solutions to create a coherent storyline for the lesson.	When sequencing students' work using technology you might sequence it in a particular order, or (since the technology makes this possible) you might choose to display multiple at a time to allow for comparing and contrasting.
Practice 5: Connecting student solutions	Identifying connections among student solutions and to the goals of the lesson that you want to bring out during the discussion.	Your role during the connecting phase is to assist students in making connections between the mathematical ideas that are reflected in their work, to help students build on each other's mathematical ideas, and to connect these ideas to your learning goal. When students' work is created using dynamic and/or connected representations it is important that this discussion includes the examination of these representations in dynamic ways to support students' connection making.

(Adapted from Smith et al., 2020)

As you can see, there are many ways that technologies, especially those with teacher platforms, can facilitate your work when selecting and sequencing students' mathematical-technological work. Next, we'll see what this looks like in practice.

An Example: Geometric and Algebraic Definitions of an Ellipse

Let's revisit Kristen's lesson arc for ellipses (Link 7.2). During the first day of the lesson, the class worked together on screens 1–10 of the activity. At the end of class, Kristen asked her students to use the GeoGebra applet linked on screen 11 (Figure 7.2) to answer the questions on screens 12–21 as homework. These questions ask about the ways in which the various parameters are related to the graph and the equation of an ellipse.

Kristen invited us into her classroom to observe her teaching this lesson. In the vignette that follows is a portion of Kristen's facilitation of a whole-class discussion on the relationships between the equation of an ellipse and the key characteristics of its graph using student responses to screens 12–21. We join her class right after they have completed a warm-up activity.

Link 7.3 Video Vignette: Kristen's Whole-Class Discussion About Equations and Graphs of Ellipses

Vignette: Kristen's Whole-Class Discussion About Equations and Graphs of Ellipses

Teacher (Kristen):	All right, so I pulled out some answers. We're gonna 1
	start with questions 12 and 13. All of your answers are 2
	so similar that I just kind of wanted to scroll through 3
	real quick. [Kristen is scrolling through the students' 4
	responses (example shown in Figure 7.3) which are dis- 5
	played for the class to see.] What does "h" do to the 6
	graph? 7

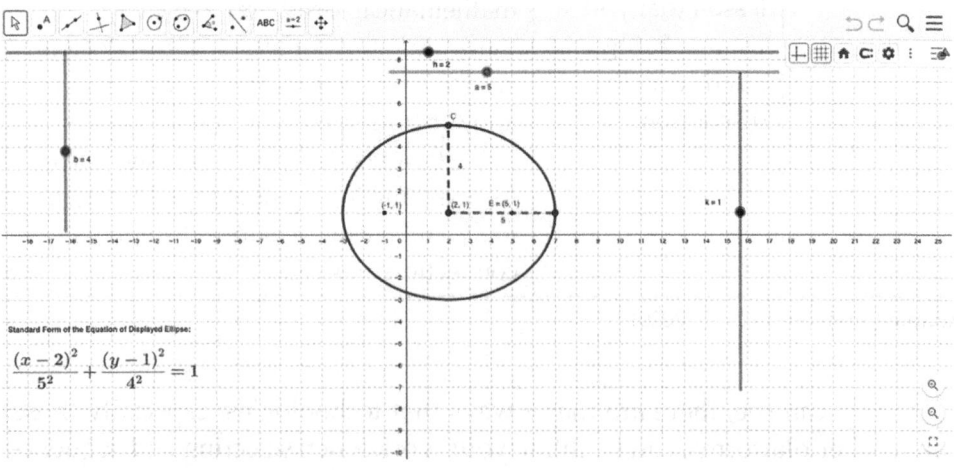

Figure 7.2 Exploring ellipses GeoGebra applet.

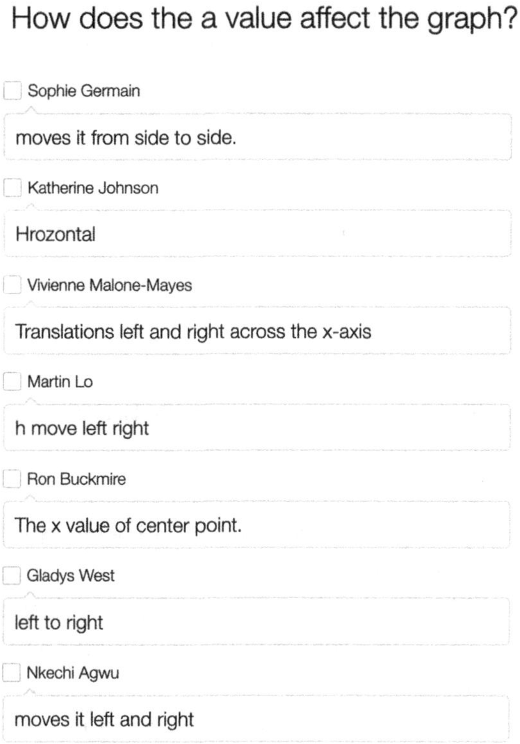

How does the a value affect the graph?

Sophie Germain

moves it from side to side.

Katherine Johnson

Hrozontal

Vivienne Malone-Mayes

Translations left and right across the x-axis

Martin Lo

h move left right

Ron Buckmire

The x value of center point.

Gladys West

left to right

Nkechi Agwu

moves it left and right

Figure 7.3 Student responses to screen 12 of the activity.

Students (chorus):	[Students talking over each other with ideas like "left and right" and "side to side".]	8 9
Teacher (Kristen):	Moves it left and right. What does "k" do to the graph?	10
Students (chorus):	Up and down.	11
Teacher (Kristen):	And that's always been the case right? Okay, so let's go to screen 14 [The prompt reads: How does the 'a' value affect the graph?]. Read through these four for me. [Kristen displays four student responses that she had selected, shown in Figure 7.4.]	12 13 14 15 16
Kristen (Teacher):	And feel free to go back to screen 9 [referring to the dynamic GeoGebra applet] and think about – are these true?	17 18 19
Student (unknown):	They're all true.	20
Teacher (Kristen):	They're all true?	21
Student (same one):	Oh no.	22
Teacher (Kristen):	They're all true? [Hands are popping up all over the room.] Oh, hands went up when you said yes. You got	23 24

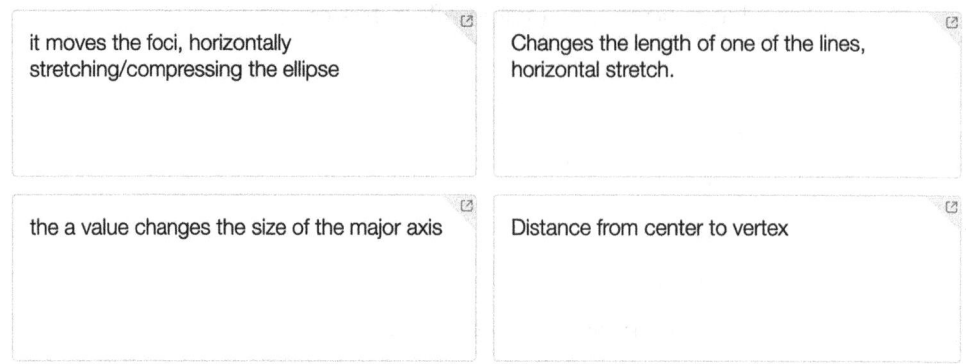

it moves the foci, horizontally stretching/compressing the ellipse

Changes the length of one of the lines, horizontal stretch.

the a value changes the size of the major axis

Distance from center to vertex

Figure 7.4 Kristen's selection of student responses on screen displayed for discussion.

	three people that were panicking and said I don't know	25
	about that Ms. Fye. […] Lotier, what are you thinking?	26
Lotier:	Okay, something for the bottom left one. I kind of for-	27
	got, I'll be real, cuz we can say it changes the size of the	28
	major axis, but technically it's not a major axis anymore.	29
Teacher (Kristen):	Okay, so interesting. When is it not a major axis any-	30
	more?	31
Student (unknown):	When it's smaller than the other axis.	32
Teacher (Kristen):	When it's smaller than the other axis. So it's not always	33
	the major axis. Is everybody with me on that?	34
Students (chorus):	Yeah.	35
Teacher (Kristen):	What would you suggest this person [pointing to the	36
	response in the bottom left] change? How would you	37
	rephrase what they said? What would you say, Sofia?	38
Sofia:	Well is it like, should it be half of the major axis?	39
Teacher (Kristen):	So let's talk about that. Let's go back and clarify on the	40
	major axis thing. [Kristen pulls up the dynamic GeoGe-	41
	bra applet.] So what were you telling me earlier, Lotier?	42
	The second Recap.	43
Lotier:	So basically I said I could. I realized that when you	44
	move "a" enough it isn't a major axis anymore. Like	45
	it does change the axis but it's not always going to be	46
	a major axis. [Kristen is using the slider to change the	47
	value of "a" as Lotier is speaking. (See Figure 7.5.)]	48
Teacher (Kristen):	Are you with him on that, Sofia, is that clear? Or are	49
	you wanting more clarity? Do you see why "major	50
	axis" might be misleading?	51

Figure 7.5 Kristen dragging the "a" slider as Lotier describes what he sees happening.

Figure 7.6 Kristen pointing to the horizontal axis as Sofia describes what she is thinking.

Sofia:	Yeah.	52
Teacher (Kristen):	Because sometimes this [pointing to the horizontal axis (See Figure 7.6)] is the what?	53 54
Sofia:	Minor axis.	55
Teacher (Kristen):	Minor axis. What direction does it always run now?	56
Students (chorus):	Horizontal.	57
Teacher (Kristen):	Horizontal, right? Why do you think "a" controls horizontal stretch and compress?	58 59
Student (unknown):	Because it's below the "h"?	60
Teacher (Kristen):	Because it's below the "h". And what do we know "h" does?	61 62
Students (chorus):	[unclear]	63
Teacher (Kristen):	What variable is also paired with the "h"?	64
Student (unknown):	x.	65

Teacher (Kristen):	"x", which also controls what movement?	66
Students (chorus):	Left and right.	67
Teacher (Kristen):	Left and right. So whatever is under the "x" is what's going to control that stretch and compress horizontally.	68 69
Teacher (Kristen):	Okay, so we've talked about how to edit maybe the bottom left one. [Referring to the displayed responses, which she has pulled back up] Anything else that we want to change? Or is there anything that we, maybe, think needs editing? Nela?	70 71 72 73 74
Nela:	So I think the top right one, it is right with the horizontal stretch, but it's not changing the length of the line, because there's not really lines in a circle.	75 76 77
Teacher (Kristen):	Mmmm. What do you guys think? What do you think they were thinking about when they said lines? Kate? [Kristen pulls the GeoGebra applet back up and uses the slider to change the value of "a" as Kate begins to speak. (See Figure 7.7.)]	78 79 80 81 82
Kate:	Because it's a stretch it is making the part in the middle longer. The axes.	83 84
Teacher (Kristen):	I think that's what they were talking about. I think they were talking about these axes. Right? Okay. And that's okay. Because these lines do exist as references inside the ellipse, even though they're not technically points that are on the ellipse. They're features of the ellipse that we can talk about. Okay, is everybody with me on that? So I'm okay with that as long as we clarify what	85 86 87 88 89 90 91

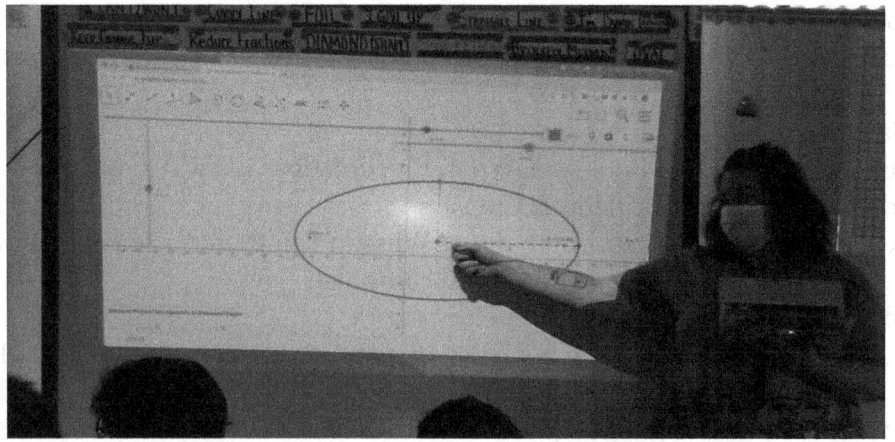

Figure 7.7 Kristen pointing to the dynamic image to annotate as Kate shares her thinking.

those lines mean. Right. Anything else that we need to   92
clarify on there? [Once it is decided that there is nothing 93
else that needs to be changed, Kristen moves to the next   94
page in the exploration and displays student responses   95
that she has selected to continue their discussion.]   96

PAUSE AND CONSIDER

1. How was Kristen's facilitation of the discussion similar to or different from how you had imagined for your classroom?
2. In what ways did Kristen use the GeoGebra applet to support student thinking during the whole-class discussion?
3. In reflecting on the Desmos/GeoGebra activity the students engaged in and the segment of the whole class discussion you saw in the video, in what ways did this activity and discussion allow students to (a) access and explore the math, (b) communicate their mathematical ideas, and (c) develop informal and powerful ways of thinking mathematically?

In this brief segment of the whole-class discussion, Kristen used carefully selected student work to guide a whole-class discussion toward a clear description of the way that changing the value of "a" affects the graph of the ellipse. She has used four students' responses that she carefully selected with the intent of problematizing some of the language used in the responses toward developing a more precise description. She began by simply asking if all four statements were true and reminded the students that they could use the GeoGebra applet as they considered the statements. This open question gave students an opportunity to reflect on not only the relationships among the sliders and the graph but also the vocabulary they had been learning. We see this in Lotier's comment about the bottom left response. When he dragged "a" in the applet, the horizontal axis was sometimes the major and sometimes the minor axis. Kristen displayed the applet and changed the value of "a" to demonstrate this change from major to minor axes. Later in the discussion, Kristen uses the applet to highlight all the dynamic representations (e.g., the equation and the graph) that are changing as she changes the value of "a" using the slider. Through this, she draws attention to the parameters and variables that all are related to the horizontal positioning of the ellipse.

The technology was key to the facilitation of the whole class discussion not only because the Desmos teacher dashboard provided a way to select and display student responses to prompt discussion, but the displayed GeoGe-bra applet also provided a way to ensure everyone was looking at the same dynamic representation as different ideas were shared, explored, and connected. In this particular discussion, student text responses were what was selected and displayed to foster discussion, but the dynamic representation was used to mediate the students' sensemaking throughout.

Using Teacher Dashboards to Support Equitable Participation in Math Discourse

Like Kristen, the Tech-Math Teachers noted how important technology has become for them in helping to create equitable participation structures in their whole class discussions. Not only does technology make it easy to share students' ideas publicly and use them to drive the discussion, but it also provides a way to include all students in nonthreatening ways. Using the anonymize feature (available on most teacher platforms) removes the stress of sharing unfinished ideas and being able to scroll through everyone's responses means that everyone's ideas are included even if they are not the few selected for the focused discussion.

- "It gives students the ability to jump in anonymously and not be afraid of making mistakes. I like to use the snapshot feature to highlight students who aren't used to participating or they're afraid that they're making mistakes or they're shy. They get to see their work on the board and realize they're doing something right and they have great things to contribute. So that is really helpful". (Michele)
- "I don't like where the only person who gets to share their response to a question is the person that gets called on. I like to create a space where everyone gets to have a voice and everybody's answer gets to be seen and/or heard. Every response can be displayed and I can scroll through them". (Samantha)
- "Being able to see student work kind of see how their thinking develops and then being able to share student work anonymously or even just being able to share the work of the quieter students who may not immediately volunteer". (Leah)

In fact, some of the Tech-Math Teachers shared that having the students' thinking so accessible helped shift their instructional strategies to more student-centered strategies.

- "This [the teacher dashboard] helped me create a more student-centered classroom where I have to do less of the talking because students have amazing ideas. To highlight those ideas to drive instruction is an extremely powerful tool". (Nick)
- "It's helped me create discourse in my classroom. I can see every student's response in real time. I can collect and sequence those responses and then that's going to kind of drive our instruction and it's all centered on those student ideas". (Nolan)
- "I can plan how I want our discussion to go because I can see what everyone is doing". (Karen)

The Tech-Math Teachers certainly make a strong case for why using teacher dashboards to make student thinking visible is so powerful. Kristen's ellipse discussion example highlighted how that could be done with students' written responses. However, in many technology-enhanced tasks, students don't just respond to prompts in writing, they also respond by creating something using the dynamic aspects of the sketches in the activity. In these situations, the teacher might carefully select student-created representations to publicly display and promote meaningful math discussions.

An Example: Exploring Measures of Center

PAUSE AND CONSIDER

Imagine your goal is for students to develop an understanding of the mean as a balance point. To that end, you have decided to use a sketch in which students can drag movable points and examine how the mean and median change accordingly as they do so (see Link 7.3 and Figure 7.8). This sketch could be used for many purposes (e.g., comparing shape to measures of center, examining effects of outliers), but since your purpose is to focus on what mean is actually measuring, you've asked students to drag the data points so that the mean and median are equal and to think about the strategies they used to do so. Take a moment to engage with the sketch and anticipate the various strategies students might use as they work on this challenge. Then consider what you might be looking for when selecting particular students' work to display with the goal of facilitating a meaningful math discussion.

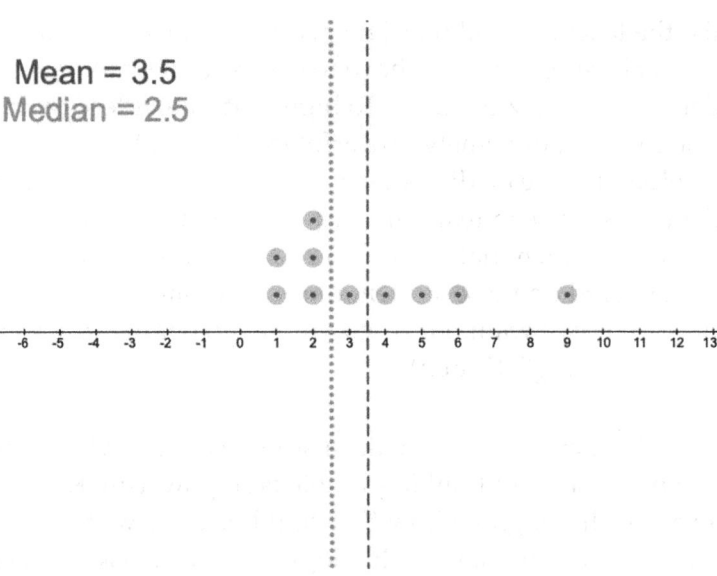

Mean = 3.5
Median = 2.5

Figure 7.8 Screen 1 of the exploring measures of center activity.

Link 7.4 Exploring Measures of Center

Nina invited us into her classroom to observe her teaching this lesson. She used this sketch with students in an Integrated Math 4 course. The students were familiar with the procedure for finding both the mean and the median but were struggling with describing what the mean actually measures (i.e., the balance point), which was important to understand before moving on to standard deviation. They were challenged to arrange the data points so that the mean and the median are equal. For this task, Nina wanted to focus on the various strategies that students used to organize the data points as they worked on the challenge. She embedded this dynamic graphing screen in a Desmos activity so that she could monitor what the students created and select and sequence student work to display to foster conversation. Figure 7.9 is a sample of the different representations the students created.

Looking across the representations Nina selected five pairs that she felt would help get the discussion rolling: Nicole and Isabella, Drew and Zaid, Jack and Robert, Jenna and Carlos, and Maddie and Hannah. Specifically, she planned to pair the graphs created by Jack and Robert and Jenna and Carlos first since they created representations in which the points were very close to (or on the mean), which was a common student approach. Next, she planned to pair Nicole and Isabella's graph with Drew and Zaid's graph as they are both symmetric, but one has points on the mean and the other does not. This provided an opportunity to discuss whether there needs to be data points at the mean and ponder what might happen if the data were not symmetric around the mean. Finally, she was going to use Maddie and Hannah's graph as it is not symmetric and the point way out to the left seems like a great opportunity to talk about how that "balances" out the location of the other points. Using the teacher dashboard presentation mode to display the

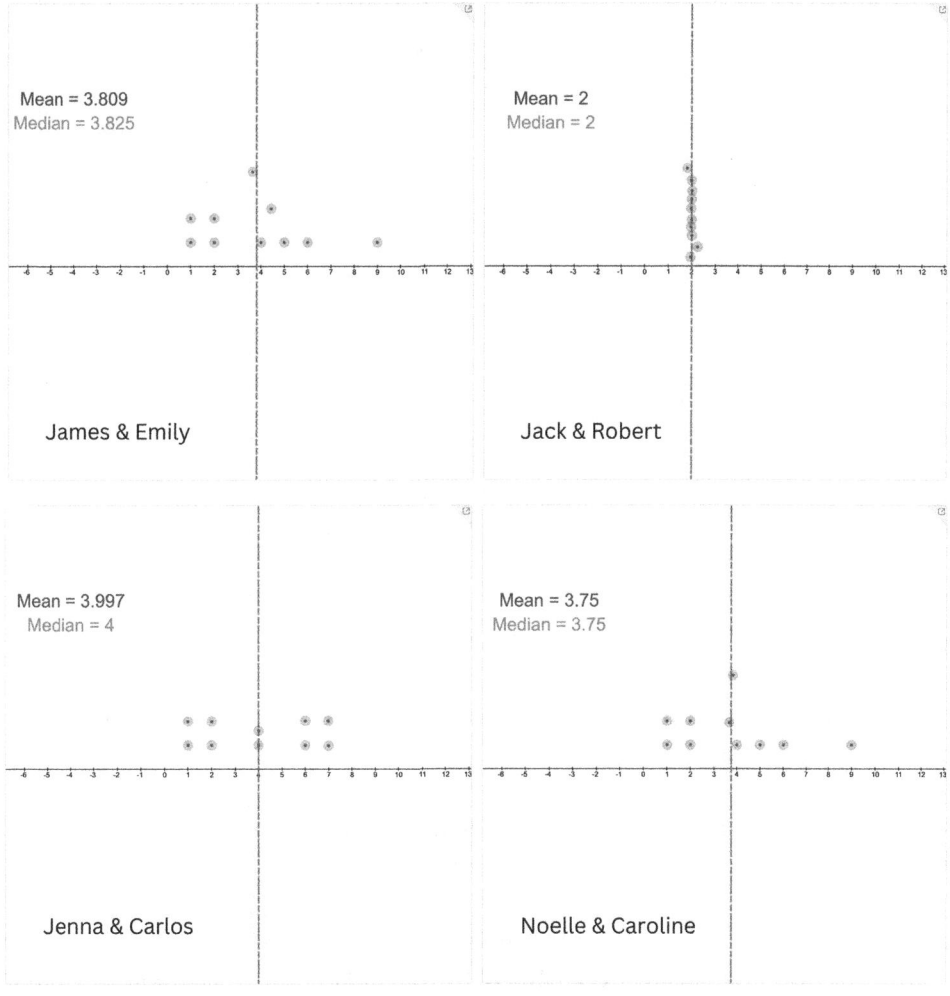

Figure 7.9 Student-created data displays on the match the mean & median challenge.

(Continued)

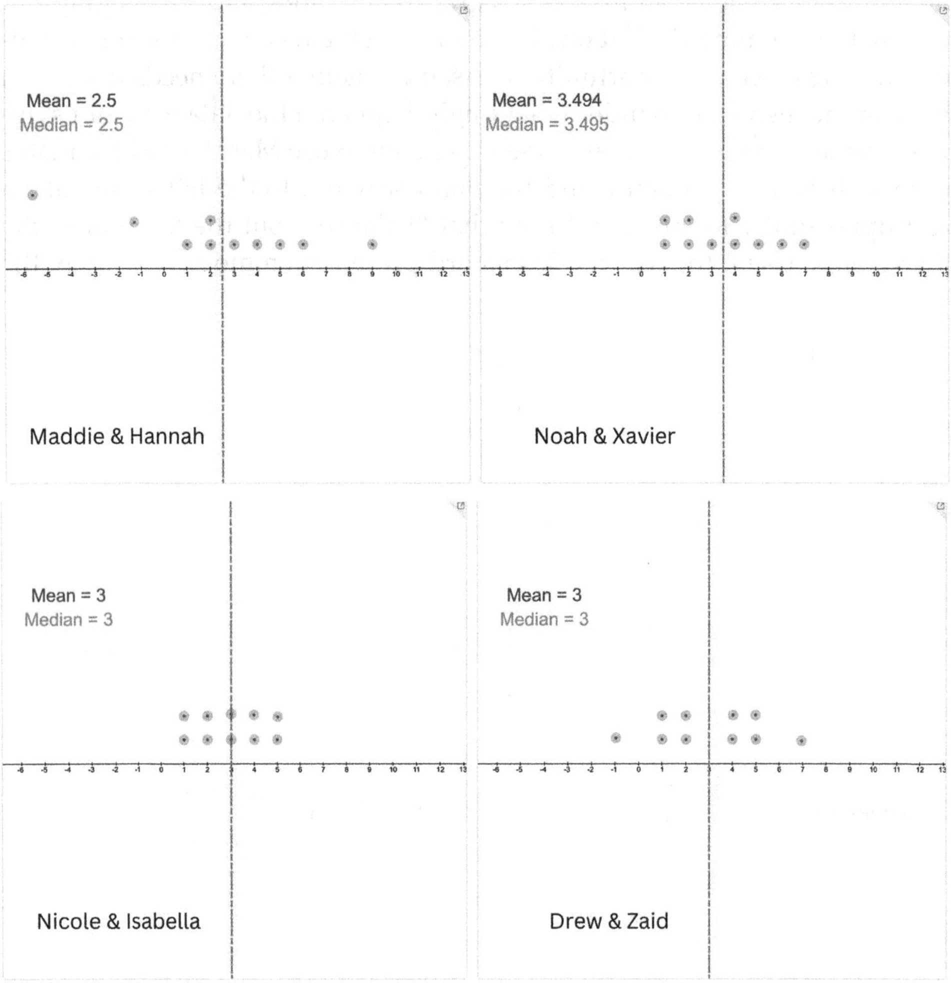

Maddie & Hannah (Mean = 2.5, Median = 2.5)

Noah & Xavier (Mean = 3.494, Median = 3.495)

Nicole & Isabella (Mean = 3, Median = 3)

Drew & Zaid (Mean = 3, Median = 3)

Figure 7.9 (Continued)

students' representations during the whole-class discussion, she facilitated the following discussion.

Vignette: Nina's Whole-Class Discussion About Mean as a Balance Point
[Note: There is no video clip for this vignette.]

Teacher (Nina):	Looking through all of your graphs I see that everyone	1
	was successful with this challenge. Very cool. Let's talk	2
	about some of the strategies you used and what you	3
	noticed about the mean and median as you explored.	4
	Here are two possible solutions [displaying Jack &	5

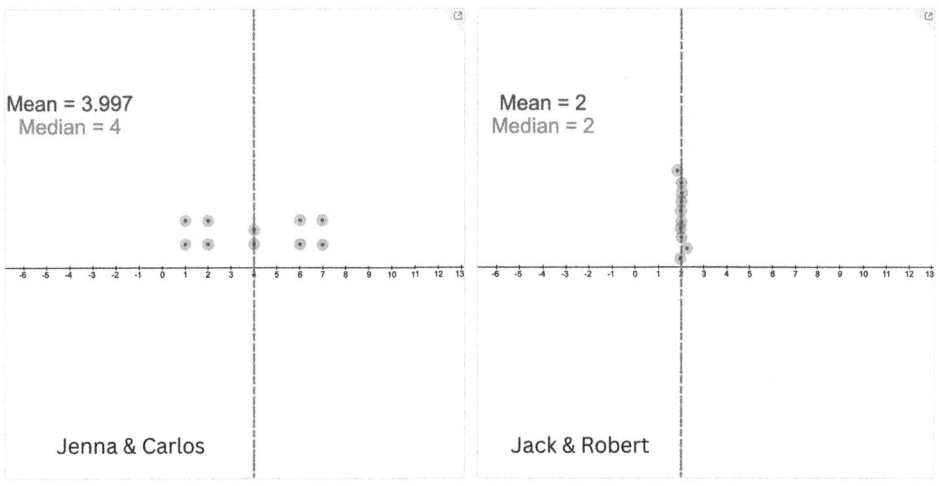

Figure 7.10 Jack and Robert's response paired with Jenna and Carlos's.

	Robert and Jenna and Carlos's graphs (Figure 7.10)].	6
	What do you notice?	7
Nicolle:	They both have points on the mean. The one on the	8
	right has all of the points on the mean.	9
Drew:	If the points aren't on the line then they are the same	10
	distance from the line.	11
Teacher (Nina):	Drew, can you say more about that? What line are you	12
	referring to? Come on up to the board if that would	13
	help.	14
Drew:	[Drew walks up to the board so he can point at the	15
	image.] The orange and purple line [pointing to the ver-	16
	tical dashed lines] the mean, which is also the median.	17
	Any of the points that aren't on the line has a pair that	18
	is on the other side of the line the same distance away.	19
Teacher (Nina):	Interesting. Can you show us what you are explaining?	20
Drew:	These two. [Using the graph on the left, Drew makes a	21
	motion from the mean to a point and then does the	22
	same to the point on the opposite side of the mean. His	23
	motion follows the arrows in Figure 7.11.]	24
Drew:	You can see it on this graph too. [Pointing to the graph	25
	on the right.] This one is a little off the line to the right,	26
	but this one is a little off the line to the right. So they	27

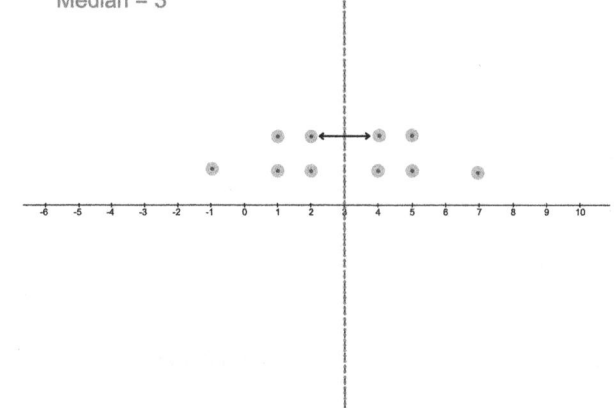

Mean = 3
Median = 3

Figure 7.11 Drew's fingers followed the arrows.

	are a pair. [Pointing to the second point from the button	28
	and the uppermost point.]	29
Teacher (Nina):	So Drew noticed that both of these solutions have data	30
	points with values that are the same as the mean, and	31
	for those that aren't, the data points seem to come in	32
	pairs that are the same distance from the mean. Does	33
	anyone have something they would like to add?	34
Caroline:	I think what Drew said about the pair of points and the	35
	distance from the mean is important because when we	36
	move just one point the mean moves too.	37
Class (chorus):	Yeah.	38
Noelle:	That drove me crazy!	39
Teacher (Nina):	Interesting … I wonder why that is. … Let's keep think-	40
	ing about that and look at some more solutions. [She	41
	displays Nicole and Isabella's graph with Drew and	42
	Zaid's. See Figure 7.12] What do you notice?	43
Isabella:	That one is ours [pointing to the graph on the left]. We	44
	started with all of the points lined up so they were all	45
	the same and so they were all the mean. But then we	46
	started to play and moved a point to the right a little,	47
	and the mean also moved right, so we had to shimmy a	48
	point the other way to keep the mean at the same place.	49
	So like, if we moved it to the right one, we had to move	50

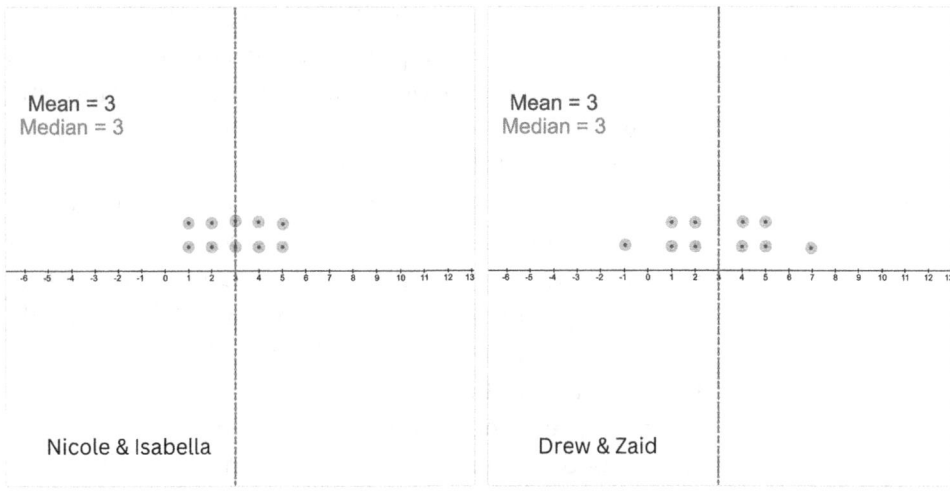

Figure 7.12 Nicole and Isabella's response paired with Drew and Zaid's.

	something else to the left one. So we ended up with it 51
	looking the same on each side of the line. 52
Teacher (Nina):	Shimmy, I like it. Your strategy description is solid. Let 53
	me restate, I think I heard you saying that when you 54
	started with your points lined up, like we saw on the 55
	last set of graphs. [The teacher quickly displays the pre- 56
	vious pair and points to the graph on the right.] When 57
	you moved a point to the left or the right the mean 58
	moved too, so you had to move a point in the opposite 59
	direction to move the mean back to where you origi- 60
	nally had it? Is that what you are saying? 61
Isabella:	Yeah. 62
Teacher (Nina):	I noticed the graph on the right doesn't have any points 63
	on the mean. How is that possible? 64
Xavier:	The mean doesn't have to have data points on it; the 65
	points just have to be spread equally around it. 66
Teacher (Nina):	Tell us more, Xavier. 67
Xavier:	[Walking up to the board to point at the graphs.] The 68
	mean isn't a data point; it is telling us something about 69
	all the points. I know to get the number for the mean I 70
	add 'em up and divide by how many I have. But here it 71
	looks like I just need to find a place where the points on 72
	the right are the same distance as the points on the left. 73
	[Xavier is pointing to the line representing the location 74

	of the mean and using his fingers to show the distance	75
	between the line and a point on one side is the same as	76
	it is for a point on the other side.] It kind of reminds me	77
	of a seesaw.	78
Teacher (Nina):	A seesaw? What do you mean?	79
Xavier:	On the playground, we can keep the seesaw balanced if	80
	we sit the same distance away from the middle. Or if a	81
	person stands right in the middle it stays balanced too.	82
	This one [pointing to the graph on the left] has people	83
	in the middle of the see-saw, this one doesn't [pointing	84
	to the graph on the right].	85
Teacher (Nina):	That's a great metaphor. Can everyone picture what	86
	Xavier is describing? Do you remember playing on a	87
	seesaw when you were a kid? [Nina quickly draws a	88
	seesaw on the board so students have a reference. See	89
	Figure 7.13.]	90
Teacher (Nina):	If Xavier is describing the kids on the seesaw as the	91
	data points, where is the mean on the seesaw?	92
Jack:	It's the middle, the balance point or the fulcrum or	93
	whatever we called it in science class.	94
Teacher (Nina):	Yeah, that is a great description. It is the balance point.	95
	Let's take a look at another. [She displays Maddie and	96
	Hannah's graph. See Figure 7.14]	97
Teacher (Nina):	This one is really different from the others, but it still	98
	works! Why? Take a few minutes to think about this	99
	with your partner. Feel free to play around by dragging	100
	the data points on your graph.	101

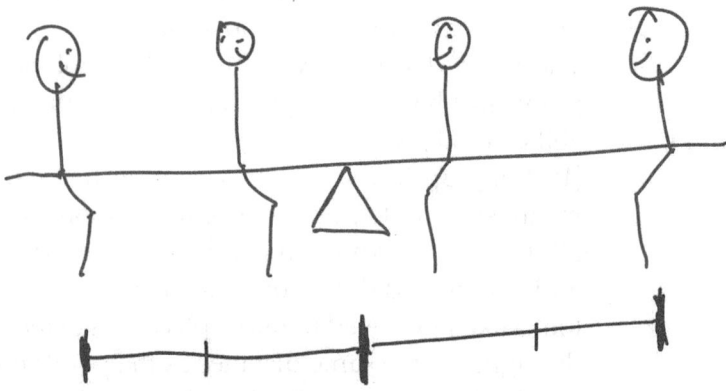

Figure 7.13 Nina's quick drawing of a seesaw on the board.

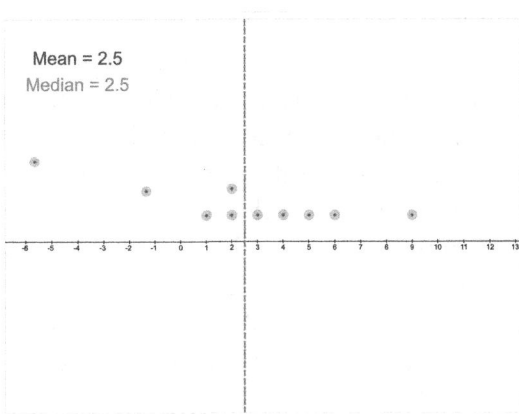

Mean = 2.5
Median = 2.5

Figure 7.14 Maddie and Hannah's response.

Teacher (Nina):	[After waiting about 3 minutes for students to talk through ideas with their partner.] Okay, what did you figure out? Why does it work?	102 103 104
Maddie:	We thought it was because each point had a pair, but it doesn't.	105 106
Teacher (Nina):	Can you give us an example?	107
Maddie:	The point at 3 has a pair that is the same distance from the mean at 2, but the point at 7 doesn't have one the same distance on the other side. But there are the same number of points on both sides.	108 108 109 109
Robert:	But the number of points shouldn't matter. That's the median, not the mean.	110 111
Teacher (Nina):	Robert brings up a great reminder. The median is measuring the value where there are an equal number of points on both sides. And since we were asked to make the mean the same as the median, it makes sense there would be the same number on each side. Let's think some more about what Maddie noticed. She said that each data point does not have a pair that is the same distance on the opposite side of the mean, some do, but not all do. So why does this still work?	112 113 114 115 116 117 118 119 120
Carlos:	I don't know if this makes sense, but it's like the points have different weights. Like on the seesaw … we need to make sure the weight is the same one each side of the seesaw for it to balance. Can I come up?	121 122 123 124

Teacher (Nina):	Of course you can.	125
Carlos:	These two balance out [pointing to the two points that	126
	are 0.5 away from each side of the mean that Maddie	127
	talked about earlier], and these two balance [pointing	128
	to the two points that are 1.5 away from each side of	129
	the mean]. I wonder if we put two on this side together	130
	[pointing to the right] they might balance with one on	131
	this side [pointing to the left].	133
Teacher (Nina):	That is a really interesting idea, Carlos. Does anyone	134
	have a suggestion for how we could test out Carlos's	135
	idea?	136
Noelle:	Find the distance each point is from the mean.	137
Teacher (Nina):	Okay. Let's try that. How far is this point from the	138
	mean? [The teacher then goes through, with the stu-	139
	dents' help, and labels the distance from the mean for	140
	each data point as shown in Figure 7.15.]	141
Teacher (Nina):	What do you notice?	142
Carlos:	They all add up!	143
Teacher (Nina):	What do you mean?	144
Carlos:	The two sides. The distances add up to the same thing.	145
	They are both 14.5!	146
Teacher (Nina):	Can anyone explain this using our seesaw metaphor?	147
Noelle:	Ohhhh, I see it now. Carlos said the seesaw should bal-	148
	ance and was talking about weight, which is like the	149

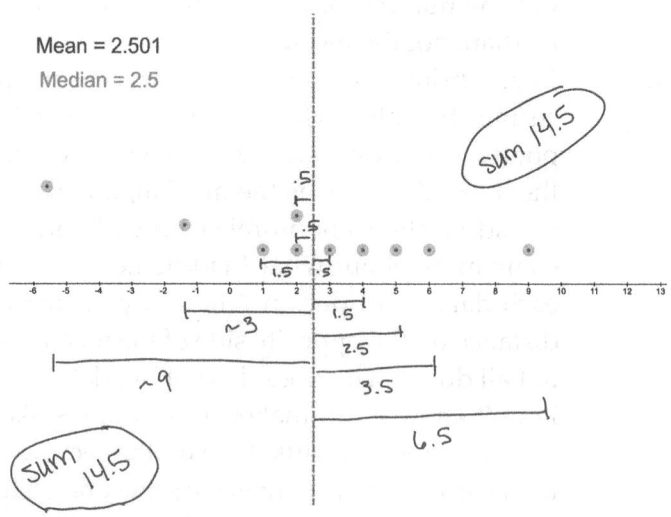

Figure 7.15 Summing the distance from the mean.

	distance here. So the number of people on each side being the same is the median, but when the weight of those people balances, or maybe where it balances, that is the mean.	150 151 152 153
Teacher (Nina):	Yeah, Noelle! It is the sum of the distances that is the same on both sides. That is pretty easy to see when the data is symmetrical about the mean, but a little harder to see here. If I wanted to keep the mean where it is, but move the point on the far left even farther left, what else would I need to do?	154 155 156 157 158 159
Tristan:	Move a point on the other side away from the mean the same amount.	160 161
Teacher (Nina):	Let's try it. [The teacher clicks on the image so that it opens in the student view and is now dynamic, the points can be dragged.] Let's drag this point [the farthest left data point] out to two more units. Tristan, what should I do to get the mean back to 2.5?	162 163 164 165 166
Tristan:	Pick a point on the right side and move it out two.	167
Teacher (Nina):	Which point?	168
Tristan:	It doesn't matter.	169
Teacher (Nina):	Hrm, why not?	170
Tristan:	Because we just need the sum of the distances to go up by two. So it doesn't matter where we get the two.	171 172

The discussion continued with Nina using the important relationship that came from the whole-class discussion to define the mean as the balance point (i.e., the point on a number line where the data distribution is balanced).

PAUSE AND CONSIDER

1. In what ways was Nina's facilitation of the discussion similar to or different from how you had imagined for your classroom?
2. In what ways did Nina use the dynamic Desmos sketch to support student thinking during the whole-class discussion?
3. In reflecting on the Desmos activity the students engaged in and the vignette of the whole-class discussion, in what ways did this activity and discussion allow students to (a) access and explore the math, (b) communicate their mathematical ideas, and (c) develop informal and powerful ways of thinking mathematically?

In this segment of the whole-class discussion, Nina used carefully selected student work to guide the class toward an understanding of mean as a "balance point", an understanding that it is crucial her students develop before she introduces standard deviation. She began by indicating she had looked at all the responses and that they were all correct. This move shifted the discussion from being about a solution, making it easy to focus on strategies instead. She used five different pairs of students' dot plots that she carefully selected and sequenced with the intent of guiding students toward the need to sum the values on both sides of the mean (i.e., examining the distribution) to see if it is balanced. She began by displaying two dot plots in which some (or all) of the data were placed on the mean as she knew this was a common strategy that the students used. She opened by simply asking, "What do you notice?" which led to Drew pointing out that everything is either on the line or equidistant from the line. This was an important insight related to understanding mean as balance. This was further emphasized when Caroline shared that the points seemed to be in pairs, moving one data point meant needing to move another as well to keep the mean equal to the median. When they then compared and contrasted the next pair of graphs, the symmetry in the graphs seemed to draw students' attention to this "point pairing" and the data being spread equally around the mean. Xavier brought up a seesaw as a metaphor for what he was noticing. Nina used this metaphor in their continued discussion, drawing a picture on the board for reference. The final dot plot that Nina shared problematized the idea that the spread of the data needed to be symmetric about the mean. She asked the students why it works and gave them time to explore it further using the Desmos sketch. It was through this exploration that Carlos and Noelle suggested that the sums of the values of the data on each side of the mean might be important. Following their lead, Nina drew the distances from each point to the mean and carried out determining the sum as was suggested.

Very similarly to Kristen's discussion of ellipses, the technology was key to the facilitation of this whole-class discussion not only because the Desmos teacher dashboard provided a way to select and display student graphs to prompt discussion but also because they were dynamic ideas that students shared could quickly be tested through dragging and observing.

Strategies for Selecting and Sequencing

Teacher dashboards often allow you to select and display student work individually or in groups of up to four. In Kristen's ellipse lesson, we saw her use four carefully selected responses in a group. In Nina's mean-as-a-balance-point lesson, we saw her use two carefully selected responses paired, as well as a single response. When deciding what to select and how to group and sequence those selections you need to not only consider how many responses to display at a time but also how you are going to select the particular responses you plan to include. Similar to Kristen's and Nina's strategies described in the prior examples, common strategies include selecting responses that have common unfinished thinking, common approaches, and different ways of expressing or representing the same idea. For example, Lauren noted,

> I really like adding one like two answers to the same group that are very different and asking students to compare them and make sense of what they are both having some value, adding value towards what the question is but from a very different perspective.

The Tech-Math Teachers also shared that taking snapshots of students' early ideas is another helpful strategy. Andrew explained:

> The other cool thing about snapshots is if you're able to be at your computer when they're working and you can actually get some of those intermediate thoughts. It [the intermediate thought] may not be there when you go back to look at it later. You're like oh that was a really good non-example, but they fixed it. I usually bring it up as this was their first thought, and that was good. I like to show some of my favorite mistakes.

Lauren seconded this idea noting that she likes "capturing enough initial thinking so we can have this meta-conversation. Look where we were, what would you tell your initial self now?" Using in-progress thinking in this way not only emphasizes the power and importance of those early ideas but also acts as a great reflection tool.

When selecting in-progress student work to display and discuss, it is important to be careful about how you pose questions and to have classroom norms about how we talk about each other's ideas. For example, if the platform has tools for anonymizing student names when sharing their responses, it is important to think about when you will use that feature and when you might not. It is also important for students to recognize that although names have been anonymized, they are talking about their classmates' ideas and that they need to do that in respectful ways. Because of this many of the Tech-Math Teachers who use such platforms regularly noted they make specific efforts to talk to students about expectations for sharing and discussing math on technology-enhanced tasks.

- "I think that in terms of norms my students know that even though it's something that they're typing on the computer and nobody might not know it's theirs, I think I've had to establish a couple times, always be respectful. Know that everybody is looking at your responses, everybody can go back to the record of it, so to always be respectful of others. Just like you would in the classroom. If you wouldn't say it to your classmate in class in person, then don't say it on the Desmos activity. I think making sure that even though you don't know the name to not laugh if you see an answer up on the board that you don't agree with. I think I try to remind them that my classroom expectations that we have in person are still at play when we engage in a Desmos activity as well". (Kristy)
- "So I really have three rules in my classroom – no more no less. The first one is that I follow all school and district policies because that is my job. The second one is that you respect yourself, your peers, your teacher, and all property located in the room. I tell them that the order is intentional. You respect yourself first, you then respect your peers, then i come last in the list because in order for them to work the way that I need them to and to do these technology tasks and to be forgiving if they make mistakes or if their peers make mistakes, that's all built on being respectful of the people around them and themselves. If they make a mistake it's not stupid. You wouldn't sit there and say that to me. You wouldn't sit there and call me stupid, that would be disrespectful. So be respectful to yourself in the same manner. (Samantha)

Selecting and Sequencing Without a Teacher Dashboard

There are some great math action technologies that do not have built-in teacher dashboards, but that should not keep you from choosing to use them. When they selected technologies that did not have teacher platforms, the Tech-Math Teachers found other ways for students to share their mathematical technological work. For example, Juan explained that his classroom is set up so that students can cast their laptops on the screen for all to see: "We are on Chromebook, so I have them take a screenshot of what's on their screen and I have them cast it to the class using my Chromecast". Nina noted that she has students take pictures of their work when using tools like Tuva or CODAP and post those pictures in Google Slides that she can then share and scroll through with the class (she can even move slides around to select

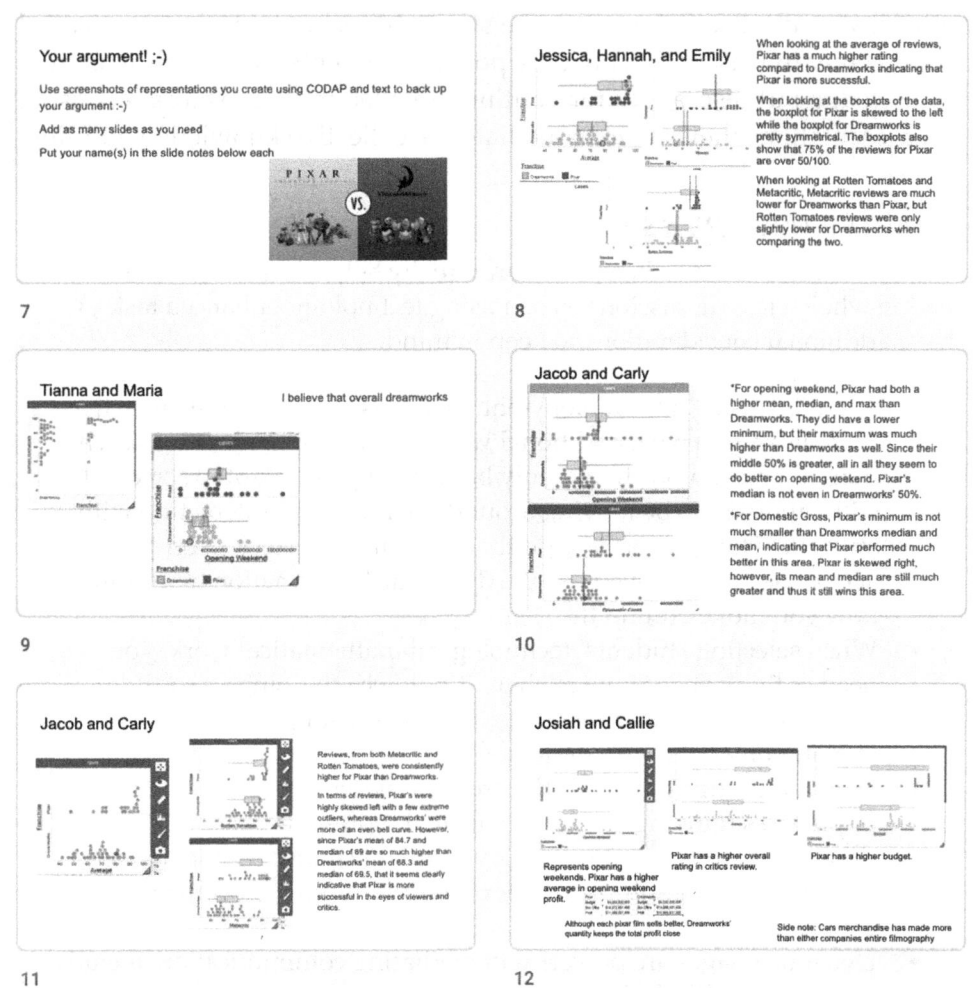

Figure 7.16 Student work created in CODAP shared in google slides.

and sequence; see Figure 7.16 for an example). If it is more helpful to the discussion to have dynamic images of student work than static, students can save their work and post the share link in a Google Doc that can then be opened on the teacher's laptop to display. This strategy works really well when students are using tools like dynamic graphing calculators, dynamic geometry, data exploration tools, or even Google Spreadsheets – any situation in which they are working with a file that they are creating content that can be saved and shared.

Whether you have the tools to screencast from students' computers or need to get creative by having them share links to, or images of, their work, whole-class math discussions that use students' technological mathematical work to center the discussion not only position students as valuable resources of knowledge but also allow for continued collaborative exploration using the representations they create to mediate the collective knowledge building.

CHAPTER TAKEAWAYS

The 5 Practices (Smith & Stein, 2011) are a really helpful framework for facilitating whole-class discussions. When using technology-enhanced tasks, we have additional considerations to keep in mind.

- If you are using a technology-enhanced task embedded in an activity builder, you have additional ways to monitor student work. This does not mean you should only be watching your laptop screen; it is still important to be moving around the room and interacting with students. However, as you do that you can also check in on students across the room or even use the dashboard information to inform how you move around the room.
- When selecting students' technological mathematical work, you need to think about how to share it publicly. In doing so consider if static or dynamic images are going to best support the ongoing exploration and discussion.
- When selecting student work consider taking snapshots of students' in-progress thinking, not just their final representations.
- When sequencing student work you might choose to display individual pieces or multiple pieces of work at a time to compare/contrast.
- Dynamic images are powerful in mediating communication, including in whole-group discussions. They give us something to show and act on to help describe what we are noticing and how different representations are connected when precise language is new or difficult to understand.

Questions to Discuss With Your Colleagues

1. If you have experience implementing the 5 Practices to facilitate whole-class discussions, what do you think will be more challenging when doing this with a technology-enhanced task? Less challenging? Why?
2. Anticipating student thinking is a critical component of the 5 Practices. How can you use this planning practice to begin to think about planning for a whole-class discussion when using a technology-enhanced task?

3. Think about a technology-enhanced task you hope to use in a lesson someday. Talk about how you might facilitate a whole-class discussion to synthesize the big ideas in the task. What student work will it be helpful to capture and use in the discussion? Is it static or dynamic work? How will you capture it so that you can easily select, sequence, and connect?

 You can find links to all the technology-enhanced tasks and supplementary videos throughout the book at https://www.tlmtresearch.com/teachingmathtechbook.

Part II

Putting It All Together

The purpose of the first seven chapters of this book was to think through the important decisions we make when including technology-enhanced math tasks in our instruction – those we make prior to the lesson and those we make during instruction. That included everything from deciding whether or not to use a technology-enhanced task, selecting specific technologies and tasks that are high quality and align with our goals, to planning an effective launch of the task, to noticing student thinking, and facilitating discussions when using technology-enhanced tasks. We hope that the Tech-Math Teachers' advice as well as the tools and frameworks we provided will help guide you as you make all those important decisions in your own practice.

In the following chapters, you will find examples of specific tasks, provided by the Tech-Math Teachers that pull everything together. Each chapter is focused on a particular mathematics domain – algebra and function, geometry, and probability and statistics. In each you will see three technology-enhanced tasks and lesson plans to accompany them. The lesson plans include learning goals, details about how to launch the task, suggestions for pacing the task, anticipations of student work, and suggested strategies for orchestrating a whole-class summary discussion. In addition, samples of authentic student work are shared with the Tech-Math Teachers' reflections on their use of the task.

Our hope is that having specific lesson plan examples from each domain will be a helpful reference as you are planning your own lessons. In addition, it provides you with some great tasks to pick up and use in your classroom!

DOI: 10.4324/9781003302285-9

At the beginning of each chapter, you will also find a list of all the technology-enhanced tasks that were referenced in Chapters 1–7 that fall in that domain. In each chapter, you have three great tasks to get you started. In short, we hope that these final chapters help you put it all together.

 You can find links to all the technology-enhanced tasks and supplementary videos throughout the book at https://www.tlmtresearch.com/teachingmathtechbook.

8

Examples of Technology-Enhanced Algebra and Function Tasks

Throughout this book, we have shared examples of tasks that are useful when teaching within the Algebra and Function domain (see Table 8.1). In this chapter, two of the Tech-Math Teachers, Juan and Dan, share and unpack their favorite technology-enhanced tasks in this domain. Juan's task is a Desmos activity that is focused on exploring the graphical representations of distance vs. time scenarios. Dan shared two tasks. The first is an introduction to polar graphs. The second is an introduction to the graph of the derivative. The purpose of including these examples is to provide you with not only some great tasks to take and use in your instruction but also insight into the thinking that goes into planning to use those tasks. For each task, Juan and Dan have included a full implementation plan as well as advice on using it in your classroom. The remainder of this chapter is dedicated to sharing these selected tasks in the domain of algebra and function.

 You can find links to all the technology-enhanced tasks and supplementary videos throughout the book at https://www.tlmtresearch.com/teachingmathtechbook.

DOI: 10.4324/9781003302285-10

Table 8.1 Algebra and Function Tasks From Prior Chapters

Chapter	Task
1	Link 1.3 "Guess My Rule" Function Machine 1.4 Introduction to Function
2	Figure 2.1 Multiple Representations of a Solution to a System of Equations Figure 2.2 Using Sliders to Explore Parameters of a Function Figure 2.9 Using CAS to Explore Factoring a Sum of Cubes Figures 2.11 & 2.12 Using CAS to Solve and Check and Equation Figure 2.13 Using DANs to Solve a Literal Equation Figure 2.14 The Three Animals Race Task Figure 2.15 Using Spreadsheets to Explore Arithmetic and Geometric Sequences and Series Link 2.4 Virtual Algebra Tiles Figure 2.24 Connecting the Unit Circle and the Graph of the Sine Function
3	Link 3.1 Building a Pen Link 3.2 Building a Bigger Field
4	Link 4.1 What Comes Next Link 4.2 Predicting Movie Tickets Link 4.3 Function Carnival
6	Link 6.5 Introduction to Vertical Asymptotes
7	Link 7.2 Ellipses (Geometric Definitions, Equations, and Graphs)

Example 1: Distance vs. Time Scenarios and the Graphs That Represent Them

This example, a Desmos activity created by the Desmos team, is inspired by a task in the Illustrative Mathematics and Open Up Resources. It is titled *Turtle Crossing*. Students explore the relationship between a graph and the movement of a turtle in several scenarios. Juan said that he likes this task because it "serves as a mental anchor for students learning about functions. These experiences are something we can come back to over and over again as we learn the requirements of a linear function". A description of the task follows.

Description of the Task

Students start the activity by being asked to draw a distance vs. time graph that is then linked to an action (a turtle crossing a beach). Once students see how their drawing is reflected in the turtle's actions, they are then asked to hypothesize the impact of differently drawn graphs on the movement of the turtle. Finally, students are asked to solve a challenge in which they need to get their turtle to cross a snake-infested beach without crossing paths with the snakes. To help support their decision, they are provided with a graph that relates the distance from the water to the time that has passed since the student pressed play. Students get to see how the graph they draw impacts the movement of the turtle as they draw it.

Link 8.1 Turtle Crossing Desmos Activity

When asked about how this task positions his students as **math explorers**, Juan explained, "My goal with this lesson is to position math as play and not work. I want students to draw and play. To wonder what will happen if I do this action, and to wonder some more". Juan's goal is that the students will explore some properties of linear functions through this play. He feels that it is important to position math as play so that students can develop some intuition about linear functions. He feels that "if you position it [math] as work you can't develop the same intuitive feel about distance and time graphs as you would letting the students play and explore". Reflecting on how the task supports students in **communicating their mathematical ideas**, Juan noted that even though

> students are not asked to communicate anything or interpret anything in the activity until screen four, since it is an open-ended, low floor task all students can begin making noticings during the launch with screen one. And by positioning the activity as play students seem more willing to share their noticings.

Finally, when Juan was asked how the task **builds on students' informal understanding** of function, he referred to the idea again of the students viewing the task as learning through play, "you're leveraging learning with fun so it gets their buy-in without them [students] thinking they are doing something for you [teacher]." By allowing students to draw anything they want for the turtle(s) to simulate, Juan feels that it can be a core experience for students to refer back to when learning about the definition of function, rate of change, and graphs of linear and piecewise functions.

TURTLE CROSSING: JUAN'S IMPLEMENTATION PLAN

Lesson Topic

Connections between scenarios and the graphs that represent them.

Learning Goals

Students will understand the relationship between simulations in context (i.e., turtles crossing a beach) and the graphs of those simulations with respect to covarying quantities.

Evidence (i.e., performance goals)

- Students will be able to sketch accurate graphs to show how distance changes over time for a moving turtle.
- Students will be able to write a story that is consistent with a given distance vs. time graph.
- Students will be able to distinguish between a distance vs. time graph that is possible or not possible to happen in a real scenario.
- Students will be able to use a distance vs. time graph to determine distance at a particular time and vice versa.

Instructional Support

This lesson allows students to experience the technology without being asked to produce any solutions before they are given a challenge to mathematicize. In this way, it is accessible to all. This lesson is available in 10 languages for students who may be more comfortable working with math in their first language. It is recommended that students work on their own devices but are placed in pairs or small groups so that they can talk with others about what they are noticing as they explore.

Prior Knowledge

Students were expected to have some initial-level experience graphing on a coordinate plane.

Essential Questions

- What is the relationship between a real-life context and a graph on the coordinate plane?
- How can I use a graph as a way to communicate the relationship between two variables?

Task Launch

Pace the activity so that students are restricted to screen 1, the warm-up, and give them a few minutes to play. If it is the first-time students have used the sketch tools, show them quickly where they are located, how they can draw and erase, and how when they have drawn something they can press play on the picture to see how their graph results in a turtle's journey across the sand. Then let them play. The prompt states: *Draw a distance vs. time graph to represent a turtle's journey across the sand. Then press play.*

As students discover new things, encourage them to share these with their neighbor and/or ask the teacher to display their work for the class to see. Some of these findings might include steeper slopes affected the speed of the turtle and negative/positive slopes caused the turtle to move forward/backward from/to the beach.

At this point just listen to students' noticings, do not dig into why they occur. They will do that as they progress through the activity. It is also suggested in the Desmos teacher moves that you select a few student graphs to display and ask, "What story does the graph tell about the turtle's journey?"

After students have shared what they have noticed and wondered on screen 1, they are ready to be launched into the activity. Simply let them know that in our exploration today, we are going to play and explore the relationships between the distance vs. time graphs we draw and the turtle's journey across the sand.

Suggestions for Pacing

During the lesson, it is suggested that you pause after screens 1, 2, and 5 to have partner conversations (e.g., What did you see? Here is what I saw.) and then hold a brief whole-group discussion.

Screen 2: On this screen, you might have students work independently for about 2 minutes individually, followed by 1 minute of sharing with their partner. As students are sharing with their partner, identify a few responses to use for a brief whole class discussion. It is suggested that you focus on two to three informal noticings (e.g., Figure 8.1) and then one that is very much connected to the context (e.g., Figure 8.2). Make sure not to share refined responses at this point so that students are all motivated to continue to engage with the task.

Screen 3: After the discussion of screen 2, open up screen 3 so that students can see the reveal of what Luca's turtle did and compare that to their descriptions.

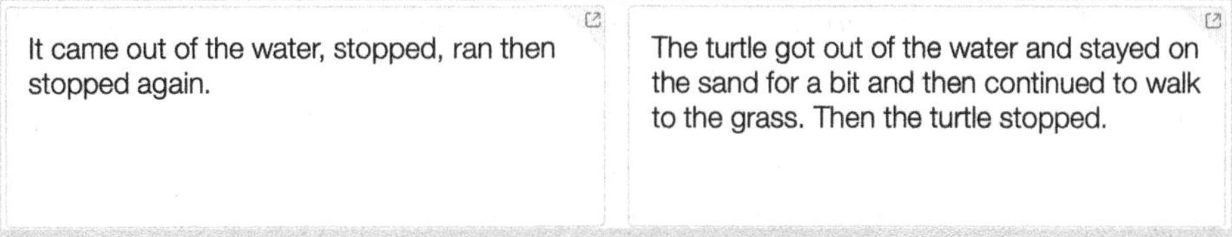

It came out of the water, stopped, ran then stopped again.	The turtle got out of the water and stayed on the sand for a bit and then continued to walk to the grass. Then the turtle stopped.

Figure 8.1 Examples of students' informal noticing on screen 2.

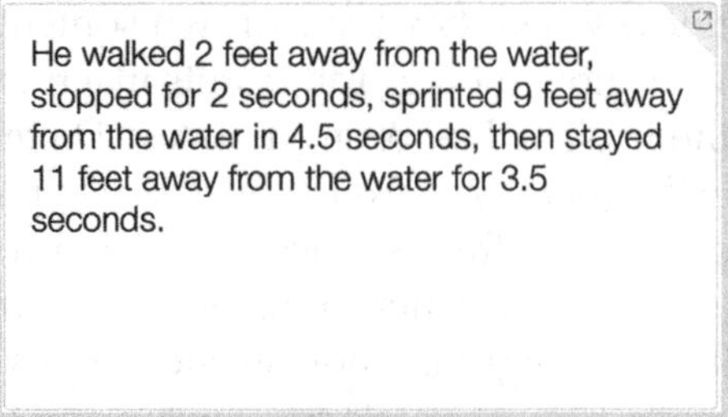

> He walked 2 feet away from the water, stopped for 2 seconds, sprinted 9 feet away from the water in 4.5 seconds, then stayed 11 feet away from the water for 3.5 seconds.

Figure 8.2 Example of a response very connected to the context.

Screens 4–5: Consider having students spend some time thinking about these two pages independently before discussing with their partner/group. After they have time on their own, have students turn and talk with their partner. Use this time to select a few student responses to share publicly for discussion. As you discuss these screens do not comment on whether student responses are "correct" or not, simply encourage justification of ideas as they relate to the graph.

Screen 6: Make sure all students experience screen 6 at the same time to discuss how the turtle "clones" and multiple turtles appear. It is important that the idea that the graph does not represent a possible journey for the turtle to come out in this discussion (i.e., what we see happening is not possible: the turtle can't clone, and it can't be in two places at the same time).

Screens 7–10: After discussing screen 6, open up 7–10. Let students know that after watching the video on screen 7 they will be drawing a graph of the turtle's journey on the next screen. Monitor these screens carefully as they will let you know if any particular students would benefit from additional support.

Whole-Class Task Summary

Screens 7–10 are helpful for facilitating a whole-class summary discussion. Begin by showing the video on screen 7 again and asking students what was tricky about graphing the turtle's journey. Most will note that it was tricky to figure out how to make the turtle go backward as this is the first example with backward movement unless they figured that out when exploring earlier screens. Follow up by asking students what they tried to make the turtle go backward that did not work and have everyone consider why those ideas did not work. Then you can move to all the aspects of this turtle's journey – forward, pause, backward, pause, forward faster all the way to the grass – asking students how to draw the graph to make each happen and why it works. After the general characteristics are discussed, be sure to discuss how to use the graph to determine specific distances at specific times and vice versa. Finally, since this activity is one that can be referenced throughout the study of graphs of functions, it will also be helpful to reflect on what features of graphs do not result in possible journeys for the turtle and why.

Advice for Teachers Wanting to Use This Task

Juan noted that this task is a great "building block" task to use early on in the school year or even on a day when the bell schedule is different because of school activities. The students are highly engaged, and it is a way to establish norms of communicating their math ideas in the classroom. Juan said, "It is an activity that you will be referring back to over and over again. It also allows me to plan discussions in future lessons that I can tie back to the students' experiences with the turtles." Reflecting on his use of this task early in the semester, Juan also noted:

> One of my takeaways from this lesson was how initial immaturity by some students quickly dissipated once students became engaged in helping the turtle solve their problem. It was insightful to see how students use their language to describe the situation. For example, one of the students described the movement of their turtle on screen 5 as "the graph glitches", which was familiar language to students who had seen particular YouTube channels, but unfamiliar to me. This allows us as a class to have a conversation about alternative language that may have a better-defined meaning, which in turn makes it an easier conversation when we follow up with the idea of function and the vertical line test for functions.

This further emphasizes how well this task is suited for supporting students as math explorers!

Example 2: Introduction to Polar Functions

This task, created in Desmos Activity Builder by Luke Walsh, and was shared by Dan. Students explore the relationship between the Cartesian and polar graphs of a function. Dan said that he likes this task because "polar graphs can be really tricky for students to wrap their head around. They [polar] break all the rules they [students] know about functions up to that point". He followed up to explain that this task allows the students to "explore and stretch their thinking" before the teacher formally introduces polar graphing. A description of the task follows.

Description of the Task

This is a Desmos activity titled *Polar Graphing*. Students use the dynamic nature of the Desmos Graphing Calculator to explore and make connections between the Cartesian graph of a function and the polar graph of the same function. They are asked to create equations to match circles of two different types. Specifically, they are prompted to explore and make connections between transformations in the Cartesian world and rotations in the polar world, as well as to explain those transformations using trigonometric identities. Then they apply what they have learned to create polar graphs to match given graphs and to create their own artwork. The activity ends with prompts that ask students to reflect on their learning.

Link 8.2 Polar Graphing!

When asked about how this task positions his students as **math explorers**, Dan explained,

> The task allows students to interact with the math that they didn't even know existed. They can play and explore things that have never been formally defined to them. And they are not just playing. They are using that play to make assumptions and anticipate and predict what is going to happen next.

Reflecting on how the task supports students in **communicating their mathematical ideas**, Dan mentioned how a lot of the communication during this task will be happening "student to student." He explained: "The task has several places where students have to explain their thinking which is great to be able to highlight different groups' thinking. But so much of the communication in this task is when students are exploring and trying to predict." For example, Dan indicated that students tend to create graphs or adapt graphs they previously created to help explain their ideas to their partners. Finally, when Dan was asked how the task **builds on students' informal understanding** of graphs of polar functions, he reiterated that "the entire task is building informal understanding" as students have no prior experiences with polar graphs and are developing their understanding as they explore. He discussed that this task is great to refer to later, while introducing formal language and explanations, because you can "refer to their experiences with the task and use their informal language".

INTRODUCTION TO POLAR FUNCTIONS: DAN'S IMPLEMENTATION PLAN

Learning Goals
- Students will understand the similarities and differences between graphing a function on the Cartesian plane and on the polar plane.

- Students will learn how to graph circles in two different ways on the polar plane – r = radius, and $r = \sin(\theta)$ or $\cos(\theta)$.
- Students will make connections between shifting functions left and right in the Cartesian plane and rotating polar functions about the origin.
- Finally, students will explore and predict the shape of different rose curves that are created by changing the frequency of the sine or cosine graph (e.g., $r = \sin(3\theta)$).

Evidence of Student Learning

- Students will be able to predict the behavior of polar graphs when given a polar equation.
- Students will create equations to match the two different kinds of circle equations using polar coordinates.
- Students will use "transformations" (which end up being rotations) of polar graphs in order to match circles.
- Students will create an algebraic "proof" using trigonometric identities in order to show the algebraic equivalence of two different trigonometric functions.
- Students will be able to match different rose curves by changing the frequency of the original circle equation.

Instructional Supports

This activity is best run with pairs of students, with one computer with internet access to share. The teacher should have a way to project the teacher dashboard to the students so that they can see the (anonymous) responses from the other students.

Prior Knowledge

Students will build on their previous knowledge of graphs of lines, transformation of functions, graphing trigonometric functions, and the use of radian angles.

Essential Questions

- How can I draw circles on the polar plane in two different ways?
- How can I rotate these circles about the origin?
- How is the Cartesian graph of a function connected to the polar graph of the function?

Task Launch

To launch this task, display the first screen of the Desmos activity, using the pacing feature to restrict students to this screen, and ask students to move a green point on the Cartesian graph and describe what happens. The goal here is for students to notice that the green dots only move up and down, and they seem to control where orange dots in the polar graph (what we call the circular grid on the right) are placed. With no other information, simply state that the goal is to move the green dots so that the orange dots match up with the open orange dots in the smiley face and to pay attention to how they make that happen. Students should be launched and ready to go at that point!

Suggestions for Task Pacing

The first nine screens are used for the students to explore how graphing a point in the polar plane (or set of points like with a function) is the same and different from graphing a point or function in the Cartesian world. With these starter screens, students are building up the ability to graph polar points and make predictions on what some functions might look like when graphed on the polar plane. Screen 9 would be a good spot for the teacher to consider pausing the activity to check in with students to see if there are similar misconceptions that have come up in their responses, or for the teacher to address student questions. For example, pairing two responses anonymously to discuss like those shown in Figure 8.3 would prompt a nice discussion.

Then share another response (see Figure 8.4 for example) and just say, "This is an interesting way to do it". Have students discuss the response

$$\left(1, \frac{\pi}{2}\right), \left(-1, \frac{\pi}{2}\right)$$

the center of the circle has a radius of 1 and the angle is 90 degrees

$$\left(1, \frac{\pi}{2}\right), \left(-1, \frac{3\pi}{2}\right)$$

You can go negative radius to get the same coordinate in the center of the circle.

Figure 8.3 Sample paired responses to screen 9.

$$\left(1, \frac{\pi}{2}\right)\left(1, \frac{5\pi}{2}\right)$$

The two different forms are the regular polar coordinate, and the regualar coordinate plus 2pi. This is because both coordinates lead you to the same point.

Figure 8.4 Another sample response to screen 9.

which should lead to a discussion about the fact that there are an infinite number of correct responses.

Other important screens to check in on are 12, 13, and 17. Once everyone has gotten through at least 13, push them all to screen 24. This is the screen that will be used to guide a whole-class summary discussion.

Whole-Class Task Summary

Once all students have gotten at least through screen 13 and have completed screen 24, use responses to this screen to orchestrate a whole-class summary discussion. Select a variety of responses on screen 24 (i.e., what students said they noticed and what surprised them) to guide your discussion of the big ideas they took away from this exploration.

Some questions to pose as you are discussing screen 24 include

- Why does a function that is a horizontal line in the Cartesian plane graph as a circle in the polar plane?
- Why does a sine or cosine graph (with a frequency of 1) end up as a circle in the polar plane?
- How do rotations about the origin work on the polar plane? How does the frequency of a sine or cosine graph influence the number of petals in the rose curve?

The goal of this discussion is to bring in some of the important vocabulary and things they have noticed so that you can continue to build on them as the unit on polar graphing progresses.

Advice for Teachers Wanting to Use This Task

When asked if he had any advice for teachers who would like to use this task, Dan said, "Don't view this task as a task that means you don't have to teach polar coordinates. It is an introductory task to build some intuition of polar graphs". Dan also discussed that it is okay to not finish the task. He mentioned that he often makes tasks with extra screens so that all students stay engaged throughout the entire class. His hope is that all students make it through screen 12 or 13. Lastly, Dan mentioned that screen 17 is a screen that he doesn't always use. The question asks students to prove algebraically a trigonometric identity. Ultimately, Dan says, this task is a great way to introduce polar graphs because polar graphs can be really tricky. They break all these assumptions about what functions look like. More concrete thinkers don't believe that this could be a function or don't understand how it could be a function. In polar coordinates you can represent a function an infinite amount of ways. So the reason for this task is to sort of explore those things and sort of stretch their thinking before you sit down and talk through it in detail. So it is perfect for exploring the ideas.

Example 3: Graphing Instantaneous Rates of Change

This example, a Desmos activity titled *Surfing the Slope*, was created by Samantha Falkner. Dan said he likes this task because "it takes a lot of conceptual thinking to understand what the derivative means", and this task allows students to build from the understanding of slope in algebra as an average rate of change to consider an instantaneous rate of change. A detailed description of the task follows.

Description of the Task
In this task, students are asked to use their prior knowledge of the connection between the slope of secant lines, the slope of tangent lines, and the difference quotient in order to predict the graph of the derivative when given a function. Students use sliders, sketches, multiple-choice questions, and text input to interact with the activity. Students are provided with the definitions of average rate of change (AROC), instantaneous rate of change (IROC), and derivative. The activity ends with a couple extension activities to challenge the students.

Link 8.3 Surfing the Slope

When asked about how this task positions his students as **math explorers**, Dan explained that this task "allows students to explore the ideas of instantaneous rate of change or sketching the graph of the associated derivative, whichever goal you are trying to develop. For example, on screen 7 the students are actually creating the derivative function without talking about what the derivative means". Reflecting on how the task supports students in **communicating their mathematical ideas**, Dan mentioned how a lot of the communication during this task will be happening "student to student." Dan says, "The task has several places where students have to explain their thinking. So much of the communication in this task is when students are exploring and trying to reason." Since students are working in pairs, their actions with the surfer and what they see happening in response to those actions are mediating their communication. You can see this happen as students say to each other "watch this" as they are explaining their thinking to each other. In addition, Dan felt that the way this task **builds on students' informal understanding** of instantaneous rate of change and sketching the graph of the derivative is powerful because

> they [students] are just using their intuition to sketch the derivative. They're interacting with sliders to create the derivative and in doing so building skills that they will use later on for what does it mean for the derivative graph to cross the x-axis or what does it mean for the graph of the derivative when the function itself is increasing.

SURFING THE SLOPE: DAN'S IMPLEMENTATION PLAN

Learning Goals
- Students will understand the connection between the graph function and the graph of the derivative function.
- Students will understand the connections between the slope of the function at specific domain values.
- Students will understand how the slope at a specific point will predict the height of the derivative function's graph at that point.

Evidence (i.e., performance goals)
- Students will be able to predict the graph of a derivative when given the original function.
- Students will be able to sketch the graph of the derivative.
- Students will be able to explain how to sketch a graph of the derivative.
- Students will be able to explain how the sign of the slope of the tangent line relates to the value of the derivative.

Instructional Support
This activity is best run with pairs of students, with one computer with internet access to share. The teacher should have a way to project the Teacher Dashboard to the students so that they can see the (anonymous) responses from the other students. In addition, we will use a Desmos activity that includes dynamic graphs (with sliders) to support student exploration of the derivative function.

Prior Knowledge

Students will build on their previous knowledge of AROC, IROC, and the difference quotient in order to sketch and interpret the graph of the derivative function.

Essential Questions

- How can I sketch the graph of the derivative when given the graph of the function?
- What does the derivative function represent in relation to the original function?

Task Launch

To launch this task, display the first screen, using the pacing feature to restrict students to this screen. Ask students to drag the slider, and answer what they notice and wonder. The goal here is for them to informally notice the difference between an AROC and an IROC. At this point, discuss their noticing and wonderings and formally introduce the vocabulary IROC and AROC. As you discuss what students notice and wonder it is important not to verify whether or not any of what they share is correct or incorrect as this could take away the opportunity for students to continue to think about those ideas as they engage with the task. Some examples of what students tend to notice and wonder are included in Figure 8.5. After this discussion they should be launched and ready to proceed to the next screens.

Suggestions for Pacing

It is suggested that you pause the activity after the students have progressed through screen 5 so that the class can analyze the cumulative answers from screens 3, 4, and 5. The teacher should ask the students if they've identified *all* possible locations, and if not, where they could add to the class's

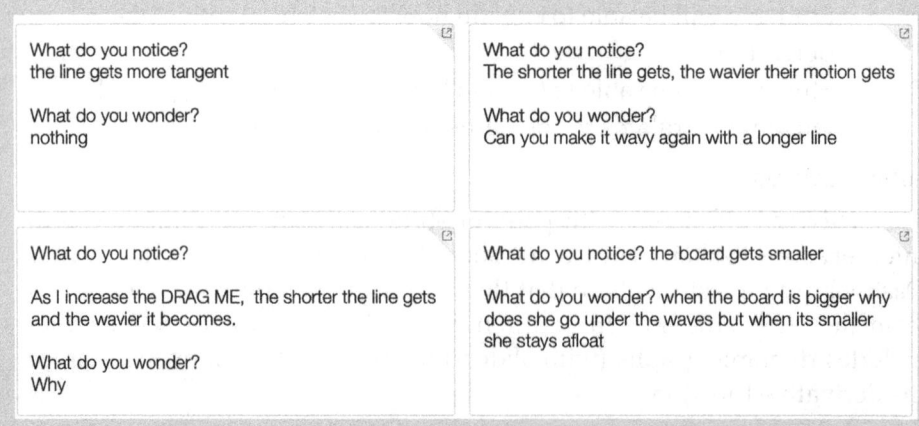

Figure 8.5 Sample student responses to screen 1.

Figure 8.6 An overlay of student responses on screen 3.

responses. For example, if you choose to overlay the classes surfers on screen 3, you might get an image like the one shown in Figure 8.6.

Notice that the students have (as a class) chosen two different zones to place the surfer. The teacher can ask the class where specifically on the graph does the zone for "good locations" begin and end? What do you notice about the beginning and ending zones for a positive slope? Have the students share their answers in pairs, and then ask/call on someone to contribute to the whole class. Repeat this for screens 4 and 5.

Screen 9 is a good place to pause and share students' descriptions of the derivative. Select responses that use the students' own language such as the two responses shown in Figure 8.7.

It is recommended that you refrain from sharing responses that are "too formal" or seem like the student Googled the answer, for example, Figure 8.8.

Whole-Class Task Summary

After all students have completed at least through screen 14, you can hold a whole-class summary discussion using student responses on this screen

a graph of the instantaneous rate of change at any given point	The first derivative represents the exact slope at every point of the original function.

Figure 8.7 Sample student responses selected to display and discuss on screen 9.

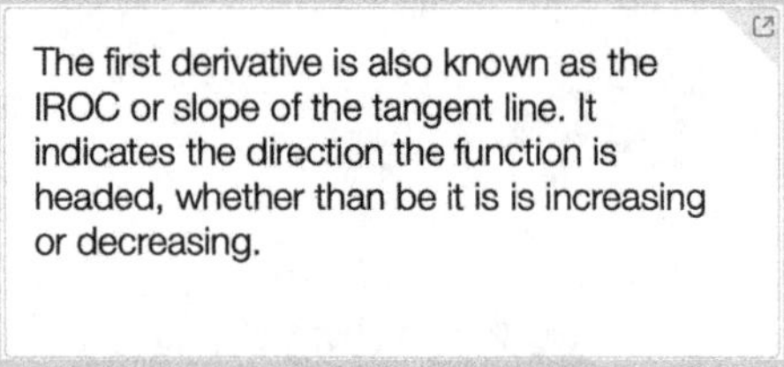

The first derivative is also known as the IROC or slope of the tangent line. It indicates the direction the function is headed, whether than be it is is increasing or decreasing.

Figure 8.8 Example of a "too formal" response for screen 9.

to drive the discussion. Select and sequence some student graphs or use the overlay feature to display all the class graphs at once (see Figure 8.9 for an example).

If you use the overlay, you might ask the class what they notice about the set of responses? What do they all have in common? They'll hopefully notice that all the sketches of the derivative touch the x-axis at the same two spots. Why those two spots? You might also ask about differences between the graphs to bring about a discussion related to intervals where the derivative should be increasing and decreasing.

Then formally discuss the connections between the graph of the function and its derivative. Specifically, what it means on the graph of the derivative when the function has a tangent line with slopes that are zero, positive, and negative.

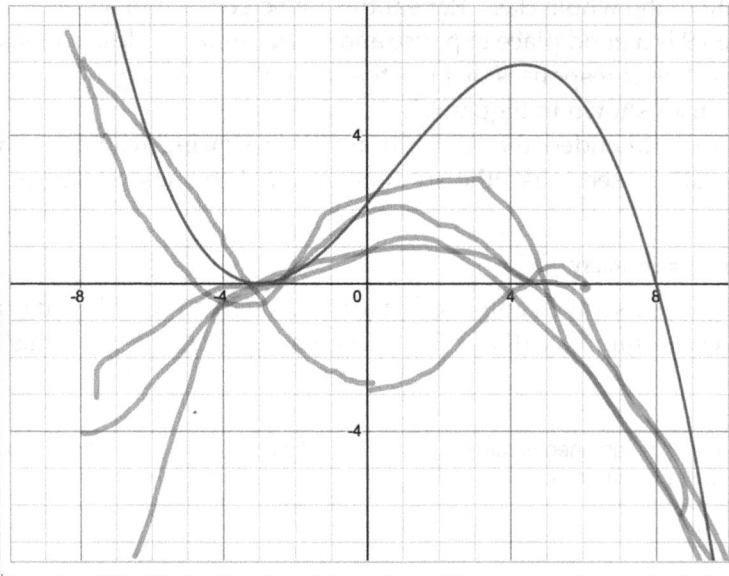

Figure 8.9 Student responses on screen 14.

Advice for Teachers Wanting to Use This Task

When asked if he had any advice for teachers who would like to use this task Dan had three pieces of advice. First, Dan said, "Remind students to use precise mathematical language regarding slope, rate of change, average rate of change, and encourage them to use that [precise language] even in their discussions with each other". Dan mentioned that even though this task is building an informal understanding it is still important to use the precise language in doing so. He also mentioned that even though screens 13 and 14 are marked as challenges, it is important that all students make it to those screens. He suggested that you make sure there's time for those screens even if the lesson carries on to a second day because "there's good stuff there, don't rush it." Finally, and most important, Dan said, "Don't try to explain what the derivative is. When they struggle that is fine. This is the very start of your unit and don't get nervous about the informal understandings at this point.".

9

Using Technology-Enhanced Tasks to Support Student Exploration of Geometry

Throughout this book, we have shared examples of tasks that are useful when teaching geometry (see Table 9.1). In this chapter, two of the Tech-Math Teachers, Andrew and Andy, share and unpack their favorite technology-enhanced tasks in this domain. Andy's task is an exploration of rigid transformations. Andrew shared two tasks. The first focuses on arc length and central angles and serves as an introduction to radian measures. The second is an introduction to geometric constructions. The purpose of including these examples is to provide you with not only some great tasks to take and use in your instruction but also insight into the thinking that goes into planning to use the tasks. For each task, they have included a full implementation plan as well as advice on using it in your classroom. The remainder of this chapter is dedicated to sharing these selected tasks in the domain of geometry.

Example 1: Exploring Rigid Transformations

This example, a Desmos Activity created by the Desmos Team that incorporates dynamic geometry tools to investigate rigid transformations, was suggested by Andy. He said that he selected this task to introduce students to the equivalence of sequences of transformations and congruence because he really liked the context of golf. Specifically, he liked that it made it into a game and was a good way for them to see how different sequences of transformations can have the same result. A description of the task follows.

DOI: 10.4324/9781003302285-11

Table 9.1 Algebra and Function Tasks from Prior Chapters

Chapter	Task
1	Link 1.1 Tangent Line to a Circle
2	Figures 2.4–2.6 The Playground Task Link 2.5 Investigating Geometric Solids Link 2.8 Conic Sections Model Link 2.9 Conic Light Simulation
4	Link 4.4 Triangle Buster Link 4.6 Volume of a Cylinder Link 4.7 Circle Chord Relationships
5	Link 5.1 Nets of Cubes Link 5.5 Arcs and Intersecting Chords
6	Link 6.1 Mystery Transformations

Description of the Task

This is a Desmos activity titled *Transformation Golf: Rigid Motion*. Students are asked to apply their knowledge of rigid transformations (translations, reflections, rotations) to complete a round of "transformation golf". Instead of hitting a golf ball, students are asked to move an "L" shape around a graph to a new location in the least number of moves or "strokes" possible.

Link 9.1 Transformation Golf: Rigid Transformations

When asked about how this task positions his students as **math explorers**, Andy explained,

> It's pretty open-ended. It gives them these transformations from the beginning, and lets them kind of play with each one to see which one might fit in which situation. This was their first big task that involved a lot of that where they had to combine things. So they certainly were exploring the various combinations.

In fact, the transformation tools that were made available on many of the activity screens provided students a way to explore and test their conjectures. Andy also noted the various ways that the task supported students in **communicating their ideas**. Not only can they show their classmates their ideas using the transformation tools, but when asked to explain their ideas, they were able to use words or pictures. Andy added,

The activity structure lends to having them at least talk about their thoughts with their group mates. I'll just kind of walk around and monitor a little bit, and try to get them doing that type of thing, talking about what they're seeing. Challenging each other a little bit about their thoughts.

He specifically noted the ways that students can use **informal vocabulary to describe their powerful ways of thinking** as one of his favorite aspects of this task.

I think in geometry vocabulary in general is kind of one of the bigger challenges. I like the way that it's open ended in the way that they can just kind of type their thoughts, or draw their thoughts. This allows them to develop some informal language and maybe some confidence and then once they realize that as long as I'm making a valid statement, I'm explaining my thinking. I feel like that's a powerful thing in a discussion. When you're having a classroom discussion is just taking what they say and being able to use that to move the class forward in the thinking, and allow them see that their thoughts are valued.

TRANSFORMATION GOLF: ANDY'S TASK IMPLEMENTATION PLAN

Learning Goals

Students will understand that there are different combinations of transformations that can produce the same image. It is the goal of this lesson to not only develop flexibility in using the different rigid transformations to produce the desired transformation but also develop the idea of efficiency of movement, fine-tuning their spatial reasoning skills.

Evidence of Student Learning (i.e., performance goals)

- Students will be able to use one or more transformations (reflection, translation, rotation) to transform a given pre-image onto an image.
- Students will be able to predict and describe a sequence of transformations that will map a given pre-image to a given image.
- Students will be able to explain why compositions of rigid transformations are not always commutative.

- Students will be able to identify and describe that a transformation is equivalent to two reflections and that a translation can be replaced with two rotations.
- Students will be able to describe the properties of a rigid transformation (i.e., that the lengths of segments and the measures of angles do not change, although the orientation and location might).

Instructional Support

This activity is best run with pairs of students, with one computer with internet access to share. The teacher should have a way to project the teacher dashboard to the students so that they can see the (anonymous) responses from the other students.

Prior Knowledge

Students will build on their existing knowledge of translations, rotations, and reflections. Specifically, how a single transformation might move an object a certain way or change an object.

Essential Questions

What transformation or sequence of transformations could be equivalent to a translation? Reflection? Rotation?

Task Launch

To launch this task, use the pacing tool to limit students to the first two screens. You might also have a picture of a mini-golf course (or individual hole) to display. Displaying the mini-golf picture, ask the class if they have ever played or seen mini golf. You will likely get a variety of yes and no responses. Then ask someone who is familiar with it to explain what mini-golf is. The student will likely say that the goal is to get the ball into a hole, but you have to work around various obstacles to make that happen. Then say that we are going to play some transformation mini-golf today and display screen 1. Here, say "The pre-image is like your golf ball, you are trying to get it to the image (or the hole) using only the transformations we have been studying, and you have to do it by avoiding hitting any obstacles!" Then show screen 2. Ask the class, "What transformation do you think might help us here?" Take a suggestion that you are sure will not work (someone will say translation – so try that!). Using the translation tool, show students that they can adjust the arrow to determine the translation direction and length (place it on a vertex of the pre-image and the image) and press "try it". The L will break as it moves through the obstacle so you can say, "Oops! That didn't work". Then challenge students to figure out what will work. Invite them to play and figure it out working with their partners. Now they are "launched".

Suggestions for Pacing

Keep the whole activity open, but once most students are past particular screens, pause and discuss on particular screens. Screens to pay particular attention to include 4, 8, 9, and 12.

Screen 4: The prompt frames the question as a challenge. This invites more participation from students who sometimes don't like to participate in discussions. On this screen, bring out the fact that you can rotate a shape in both directions. Ask students to draw what they think would happen to the shape if it is reflected. Then do the same with a rotation. Can you place the line of reflection anywhere that might make the pre-image line up in the desired location? Is there anywhere you can put the center of rotation to make the pre-image line up with the desired location? You might have a student come to the front and draw on the whiteboard where the screen is projected (or just show their screen if they were able to draw effectively).

Screen 8: Have students debate for a little bit, hearing their reasoning and letting them hash it out on their own if possible. But to bring out the point, ask, "If both transformations were the same, and only the order is changed, then does the result *always* end in the same location?"

Screen 9: Have students discuss their ideas on what transformations might work for this challenge. Have them draw, visualize, and discuss without actually testing it out (see Figure 9.1). Bring out which transformations

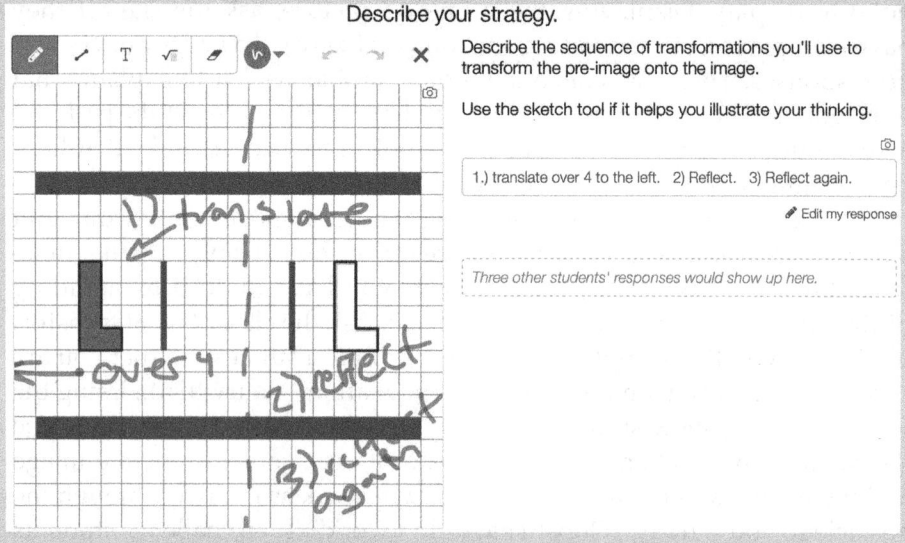

Figure 9.1 Sample response for screen 9.

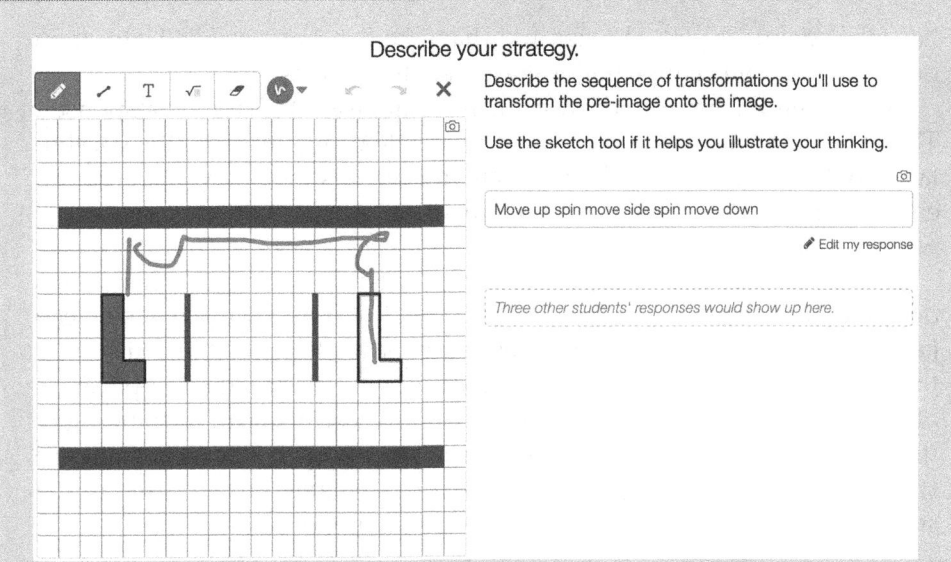

Figure 9.2 Sample response for screen 9 that lacks detail.

or combinations of transformations might be the most efficient as opposed to ones that just work. However, if they are immediately going to the most efficient route, you might encourage them to find other less efficient ways to foster some flexibility and creativity. When selecting responses to display to foster discussion consider displaying those that are missing detail in the drawing or description (see Figure 9.2 for an example), as they provide a great opportunity to compare and contrast strategies and practice describing them precisely using the language of transformations.

Screen 12: Have students draw or show their justification for their answers. Having them do this in front of the class helps other students see it if they are having trouble visualizing it. It also makes them see that math often has different paths to the right answer and gives me an opportunity to celebrate those different paths.

Whole-Class Task Summary

Use screens 16 and 17 to facilitate a whole-class discussion to summarize the big ideas of the lesson. Screen 16 provides an opportunity to use the vocabulary precisely. Screen 17 is a great opportunity for debate with students creating arguments and critiquing the arguments of others.

In addition to realizing that a transformation might be equivalent to another sequence of transformations, the big goal is that by the end of this discussion, students understand that this particular set of transformations (i.e., rigid transformations) results in congruent figures (even if we don't have that language yet). The size and shape do not change; only the location or orientation will change under these transformations.

On screen 16, there often is not much variability in the student responses. The variability that occurs is typically related to being unsure of the vocabulary, not the ideas. Ask students to tell you what is meant by position, orientation. Those are usually the terms that they confuse or just don't know what they mean.

On screen 17, just like on many of the other screens, have students debate each statement and have them argue it out and see if one side can justify their answers. Encourage them to draw or show the other side why they think the way they do. Having the students discuss their application of each of these different transformations really tests their understanding of them collectively and develops their spatial thinking. If time permits, have students use the technology (could use the last screen) to test each of the statements and see if they appear to always be true or false.

Advice for Teachers Wanting to Use This Task

Andy noted that this task works really well for students who have some experience with rigid transformations, but he would not use it to introduce those transformations. It is more focused on making sense of the possible outcomes of a series of rigid transformations and sets up congruence really well. Even if students have not played golf, it is very intuitive so they pick up on it quickly. In addition, Andy shared that this is a great activity to help kids get used to debating. He explained:

I think it could be the first activity that you kind of use to help kids learn how to debate and how to talk respectfully about somebody else's work. I think this is a pretty open-ended thing that kids can kind of share their opinion in a safe place without having to worry about not feeling like it's not technical enough or whatever so depending on, you know I guess your pace you're pacing guide to what you want to use. I think that this could be something where you start trying to teach kids how to have a good class discussion, and how to debate, and how to critique each other's drawings, and thoughts and things in a respectful way and other than that.

Example 2: Introduction to Radians

This is a Desmos activity titled *Pizza Slices*, created by Luke Walsh, adapted by Andrew. Andrew said he likes this task to introduce radians because it begins as more of a game comparing pizza slices. This is a great, informal way to have students begin to compare portions of circles without using any specific or formal language. As the lesson develops, students describe their difficulties in answering the questions and are led to want more specific information in order to make a better 'guess'.

A description of the task follows.

Description of the Task

Students are asked to investigate the relationships of "slices" of pizzas by attending to their sectors, arc lengths, and central angles. Students will start with a concrete example of "slices" of pizza and work to the more abstract idea of what a radian is and that sectors of circles can be described using central angles and arc lengths measured using degree and radian measures. The overarching goal of the activity is to build an understanding of the relationship between degree and radian measurement. Students are then introduced to formal ideas of how parts of circles can be described and measured. This leads students to formalize the idea of a radian being the radius-unit of any circle. Finally, students are introduced to the connection between degrees and radians and practice converting from one measurement to another.

Link 9.2 Pizza Slices Desmos Activity

When asked about how this task positions his students as **math explorers**, Andrew explained that this task allows students to "bring in an abstract concept of parts of a circle into a context they know and feel comfortable exploring and discussing". This happens as students explore radians through the "gamification of pizza" to develop an understanding of a radian. Students have opportunities to use sliders to measure angles and arcs using degrees and radians as they explore. Andrew noted various ways that the task supported students in **communicating their ideas**. He explained that the context of pizza slices, something students are familiar with, provided a concrete example to use when sharing their ideas about angle and arc measures. As students used the various dynamic tools to explore, they also used terms and phrases like *crust* and "edge of a slice" as they shared their ideas with one another. Andrew also pointed out that this task incorporates a lot of different Desmos components (e.g., answer boxes, multiple choice, and card sort) that provided students different ways to communicate their ideas. When considering the ways that students develop **informal and powerful ways of thinking** when engaging with this task, Andrew shared that he really likes the way that the first eight screens develop the students' curiosity about what math they are exploring. He mentioned that students often say to him that "they are trying to determine what math they are studying by guessing the pizza slices." Andrew feels that this informal play allows them to start developing the purpose of the task before he even tells them. Additionally, this task allows students to visualize the radian wrapping around the circle. This provides students "so much intuitiveness" by developing an understanding of radians in this way. Andrew mentioned he's seen other hands-on tasks demonstrate this, but he believes this is one of the most powerful parts of the task.

PIZZA SLICES: ANDREW'S TASK IMPLEMENTATION PLAN

Learning Goals

- Students will understand that there are multiple ways to measure angles of a circle.
- Students will understand the different parts of a circle (sectors/arcs).
- Students will understand the connection between degrees, radians, and the circles they measure. Specifically, they will understand how any circle (or part of a circle) can be described with 360 degrees, or 2π radians.
- Students will understand the difference between degree and radian measurements and be able to convert from degrees to radians and vice versa.

Evidence (i.e., performance goals)

- Students will be able to explain how a "slice" (sector/minor arc) of a circle can be measured by the central angle or arc length (portion of circumference).
- Students will be able to identify equivalent arc lengths and central angles.
- Students will be able to convert between radians and degrees.
- Students will be able to explain how a radian is a measure of a circle using the circle's radius as a unit.
- Students will be able to represent a radian measure greater than 2π in a circle.

Instructional Support

This activity is best run with pairs of students, with one computer with internet access to share. The teacher should have a way to project the Teacher Dashboard to the students so that they can see the (anonymous) responses from the other students.

Prior Knowledge

Students will build on their knowledge of the parts of circles (especially that a circle is composed of a central angle of 360 degrees, the radius/circumference size changes with the size of the circle) and parts of circles (sectors, arc length, etc.).

Essential Questions

- How can I describe how much of a circle I have?
- How can I describe a circle in different ways (interior angle vs. outside circumference size)?
- What are radians, and why are they helpful when describing circles?

Task Launch

Start the task using the pacing features to restrict students to screen 1 and have the activity paused so they can't begin yet. Tell students you are going to play a guessing game. Then show screen one and explain the prompt: You will see a picture of a slice of pizza. Then you have to figure out how many slices of that size would make up a full pizza. Unpause and let them all answer the first screen. They will make their selection, and then the answer is revealed. Once everyone has responded, use the teacher dashboard to show them the frequency display of how the class responded overall. Then explain that the next few screens will be similar, and challenge them to be as accurate as possible. (Note: Make sure they know that points don't matter. They should do their best but not worry if they are wrong.) Then open up screens 1–8, and the task has been launched.

Suggestions for Pacing

It is suggested to use the pacing feature in Desmos to help students progress through the screens at a specific time. This also allows time to bring the class together for brief consolidation of ideas before moving on to the more formal parts of the activity.

Screens 1–8: First, "pace" students into screens 1–8. This is the informal part of the lesson and allows students to play with the idea of "slices" of pizza and has them reflect on what could have gone better on screen 8. After students answer the prompt on screen 8, note that even a simple "guessing game" can be difficult. Use the snapshot tool to select and display some student responses to point out some of the difficulties students highlighted. First, display those that are thinking only about the context (i.e., pizza); for examples, see Figure 9.3. Then display a few responses that indicate students are beginning to think about "what is the math in all of this". For example, see Figure 9.4. Then highlight that *everyone* struggled with finding

> The difficulties I had were... imagining the whole pizza

> The difficulties I had were underestimating the actual size of the slice

Figure 9.3 Sample responses to screen 8 focused on the pizza context.

> The difficulties I had were estimating how much space the size of the slice would take up

> The difficulties I had were... not being to know the exact degrees or much about the slices.

Figure 9.4 Sample responses to screen 8 that allude to the math.

> The crust length.
> _____
> Crust length is a good clue because... I measured the crust with my finger and went all the way around

> The crust length.
> _____
> Crust length is a good clue becuase you can compare it to 360

> The angle measure measurement.
> _____
> Angle measurement is a good clue becuase... i could subtract that ffrom 360

> The circumference of the pizza.
> _____
> The circumference of the pizza is a good clue because... it is the length around the entire pizza

Figure 9.5 Sample responses for screen 9.

the pizza slice size, maybe with some more tools we can guess a bit better. This leads into screen 9's prompt, so open up screen 9.

Screen 9: Give students a moment to respond and then display the frequency of their choices. It is also recommended that you use snapshots to show some justifications students provided for each of the options (see Figure 9.5).

Screens 13 and 14: It is helpful to come together at the beginning of screen 13 to explicitly discuss how there is crust length and angles that we

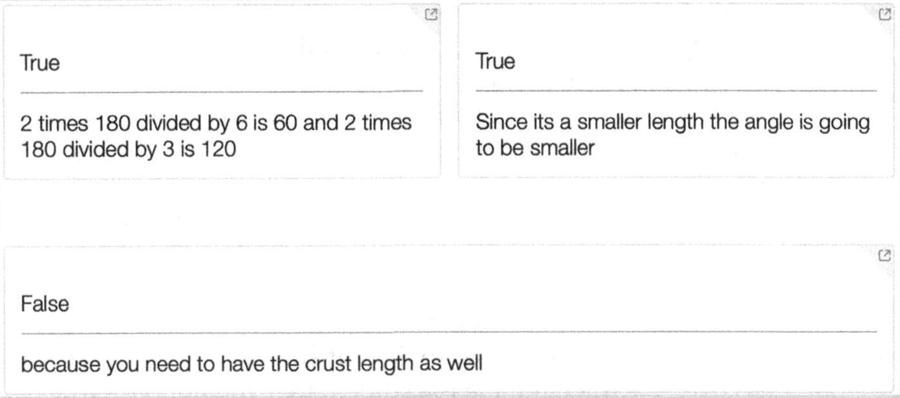

True

2 times 180 divided by 6 is 60 and 2 times 180 divided by 3 is 120

True

Since its a smaller length the angle is going to be smaller

False

because you need to have the crust length as well

Figure 9.6 Sample responses to screen 14.

can match up with the slices. Clarify that the inequalities show a range and that we are looking for the pizza slice(s) that fit in each range. As students are working, keep an eye on the card-sort correctness; the teacher should view and look for groups or individuals that have a difficulty sorting the cards. Often, students have trouble thinking of pizza in terms of crust length, the angle distinctions seem to go better for them. It is suggested to have the activity pacing open through screen 14 so students that finish the card sort can move on to the reflection prompt on 14.

Once all students have completed the card sort it is helpful to then go over it as a whole class as well as troubles students noted on screen 14. Not all students get to answer 14, but try to show the two stances – both true and false – and use those to prompt discussion about the relationship between angle and arc length as well as radians and degrees. See Figure 9.6 for some sample responses.

Screens 15–20: These screens can be opened up depending on time to be more individual or group work. As most groups are finishing, move to the whole-group summary, but consider leaving screen 20 open so that students that don't get a chance to talk in the whole-class summary discussion still can provide their input.

Whole-Class Task Summary

When doing the whole-class summary, it is helpful to run through the whole lesson and show how it is informal → formal thinking. Then be explicit about what formal ideas were covered.

Specific screens to show in the whole-class summary are the following:

- Screen 8, where you can show the pizzas with different numbers of slices

> - Screen 13 to bring the order of the screens we just saw on screen 8 back into the focus of ranges of crust length (which we now call radians) and angles (degrees)
> - Screen 15 to describe the crust wrapping around the pizza as you move the slider to show again how crust length (arc length) is a radian
> - Screen 19 to refresh how we can convert/compare radians and degrees
>
> This is also time to use snapshots again to show specific student thinking or even just pull out a response or two of poignant student answers that may have not been covered earlier. If there is a student that you noticed throughout the lesson who has made some great mental jumps and is descriptive of the whole class's thought process, it is helpful to follow that students' responses throughout the activity so the whole class can see the informal → formal path we took as a whole.

Advice for Teachers Wanting to Use This Task

Andrew's biggest advice is to make sure you work through the entire task and try to think through it like the students: "You need to know what goes into answering each portion of each question". Andrew noted that this is really important to him because he finds it "difficult to think of pizza slices by crust" and that he has to "think through it to be able to explain it to students". He also mentioned keeping the pacing slow. "Don't open up additional screens because some students will speed through. Just open up 2–3 screens at a time so that they see they can work ahead but still be engaged".

Example 3: Geometric Constructions in GeoGebra

This example, a GeoGebra Activity, was created by Andrew. Andrew likes to use this task to teach about geometric constructions because "it allows students who can't draw super well on paper with a compass and straightedge to draw a perfect circle each time and make sure all the points line up". Also, unlike paper-and-pencil constructions where your construction won't always fit on the page, Andrew prefers to use GeoGebra because of the "infinite work space" and because students can "make the constructions bigger or smaller through dragging" to represent all possible examples. A description of the task follows.

Description of the Task

This is a 3-day (1.5 hours/day) series of activities embedded in a GeoGebra Book titled *Geometric Constructions*, created by Andrew was adapted from the work of Ku, Yin Bon (Albert), John Golden, and Jonathan Garfield. Students work through the activities to complete 13 basic constructions with a compass and straightedge. Each construction builds on previous constructions (skills learned), and each set of constructions has a group of construction challenges at the end to extend a student's newly acquired knowledge.

Link 9.3 Geometric Constructions

When asked about how this task positions his students as **math explorers**, Andrew shared that "it gives them that chance to explore it [constructions] and supports the students with the tools [compass and straightedge] because they can play around with using those tools together and explore all the properties of circles and lines". Andrew also noted the various ways that the task supported students in **communicating their ideas**. As students work together to complete the construction challenges, they are also challenged to determine the properties of their constructions, such as congruence and parallel / perpendicularity. Students use the dynamic capabilities of the construction tools to annotate as they communicate their thinking about these properties as well as the soundness of their constructions. In addition, GeoGebra Classroom allows Andrew to see all the students' constructions in an instant. This means that he can "elevate students' thinking and understanding" by using their work during class discussions. Finally, Andrew discussed how students develop **informal and powerful ways of thinking** about geometric constructions in that "the task is designed so that each construction builds on itself. The students can start using these basic constructions with each other to create more detailed constructions". Since students are using dynamic geometry tools rather than a straightedge and compass, students can drag and try to "break" their constructions allowing them to play and notice properties as they are still formalizing their thinking about the differences between sketching and constructing. Finally, Andrew mentioned that the task is "very powerful because it allows the us [students and teachers] to do what we can't do on paper. On paper we can't show a thousand examples in a very short amount of time".

GEOMETRIC CONSTRUCTIONS: ANDREW'S TASK IMPLEMENTATION PLAN

Learning Goals

Students will understand how several constructions can be created using only a compass and a straightedge. Specifically, students will work with the following constructions:

- Duplicating a Given Line Segment
- Duplicating a Given Angle
- Creating Equilateral Triangle
- Creating a Perpendicular Bisector
- Creating a Perpendicular Line Through a Point on a Line
- Creating a Parallel Line Through a Given Point
- Creating an Angle Bisector
- Finding the Circumcenter, Incenter, Centroid of a Triangle
- Finding the Center of a Given Circle

Evidence (i.e., performance goals)

- Students will be able to use a compass and a straightedge to create the constructions outlined in the learning goals.
- Students will be able to explain the steps (in writing/verbally) of how they created each construction.
- Students will be able to apply their understanding of constructions by using previous construction techniques to create more intricate constructions.

Instructional Support

Each student (or pair of students) needs a computer with internet access. Within GeoGebra, only specific tools are given so that students accomplish the task with a compass/straightedge and not be overwhelmed with the variety of tools available on GeoGebra. Students may find it helpful to have paper or a small whiteboard to sketch out ideas.

Prior Knowledge

Students will build on their knowledge of a compass and straightedge in order to find methods to construct specific geometric properties of shapes, such as the center of a circle, copying line segments, and so on. Students should have a basic understanding of the geometric definitions of *line*, *ray*, *segment*, *circle*, and *arc*.

Essential Questions

- How can I use a compass and a straightedge to create geometric constructions?
- What geometric objects can we create with a compass and a straightedge?
- What specific things/parts/locations of geometric objects can be found using only a compass and a straightedge?
- How can we combine construction techniques to create new/more complex constructions?

Task Launch

Begin by explaining the difference between constructing and sketching. A good way to do this is to provide students with a GeoGebra file that has two figures on it, one that "breaks" when it is dragged and another that does not (see Link 9.4 for an example using squares). A construction is something that is built using tools in accordance with the properties that define the object. In geometry, the tools of construction have historically been a straightedge and a compass.

Link 9.4 Sketch or Construction?

Explain that constructions allow us to see how much we can accomplish with reasoning and these "basic" tools. From only a compass and straightedge you can create complex geometric objects as well as find several parts of other objects. Constructions allow us to "see" complex mathematical ideas and logical deductions without the use of numbers, formulas, algebra, and so on. So our goal over the next few days is to use constructions to explore these relationships.

Next give students the link to the GeoGebra Book (Link 9.3). If this is the first time that they are using a GeoGebra book, show them how to navigate the book as well as the activities within the book in the left side pane and the "previous" and "next" arrows at the bottom. Then use the first book, *Compass and Straightedge*, to display and define what each tool is and what it can do. Then have students open the third activity in the book, "Straightedge and Compass". Show them where to find the straightedge and compass tools, as well as how they work. It is also helpful to show them how to use the text tool and the undo tool, move to orient the plane, and how to refresh the space to return it to its original settings (this will erase everything). Then allow them time to play around with the tools for a bit. Consider sharing a few things they have created using the tools so they can see how you will be able to select and display their work during discussions as well. At this point, they should be ready to begin working on the constructions.

Suggestions for Pacing

The overall activity is set up in three 1-day parts. The beginning of each lesson will be teacher-guided allowing students to see specific geometric constructions. Students will then be able to practice the constructions, apply the constructions to new tasks, and, finally, participate in challenges that apply the basic construction in new ways. Constructions 1–4 and the associated practice and challenges are for day 1, with constructions 5–8 on day 2, and 9–13 on day 3. Next is a suggested pacing for day one, the same structure can be used for each of the 3 days of instruction.

For Day 1 – Constructions 1–4: These are the building block constructions, which will be used in later constructions. In GeoGebra, there is no pacing feature to restrict student movement through pages, so it is important to give students directions on which page to be on.

Have students work on the first two constructions, "Duplicate a Line" and "Copy an Angle". At the top of each page, there are written instructions on which tools to use and how to check. However, it does not explicitly tell them how to do it. Since this is their first introduction to constructions, most students will try to use the line tool and the measuring tool to copy the line and the angle; thus, they will not be exact (see Figure 9.7).

Next, come together as a class and talk about the fact that although we have measurement tools that are fairly accurate, we want to figure out an "exact" way to duplicate the line and copy the angle. If there are some good examples of "close but not exact", show the anonymized student work of

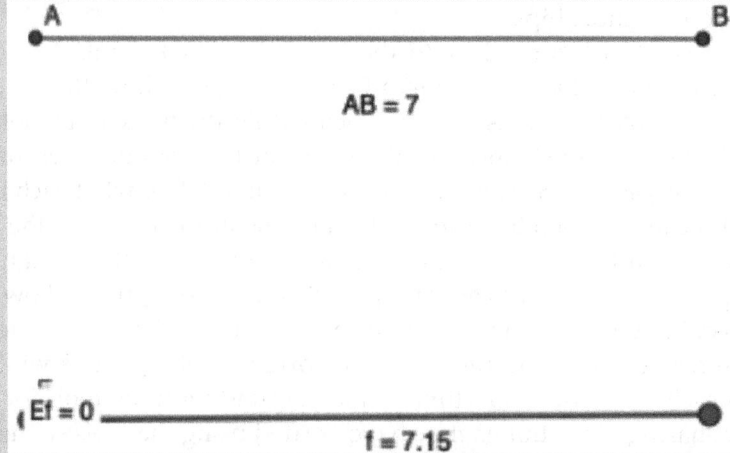

Figure 9.7 Sample first attempt to duplicate the line segment.

examples and non-examples and celebrate various strategies students used. Make sure to also talk about accuracy and what is "close enough". After noting that we need our duplications to be exact (i.e., congruent to the original), lead the discussion to include the properties of a circle and how using a modern compass can copy an exact length, since all radii of a circle are congruent. It is also important to talk about duplicating a line and if a line has a different orientation, is it duplicated or not.

Then have students go back and practice using the tools to complete constructions 1 and 2. Sample responses are shown in Figures 9.8 and 9.9.

After working through all four constructions in a similar fashion we pause for a whole-class summary. Then students work in small groups/individually to work on the construction practices and challenges (1–4). All of these are differentiated to add enough complexity so that students can explore an idea further, or if they are not yet confident in the construction, practice the steps we worked through together.

There is some awesome student thinking that comes from the practice and challenge problems. Plan to start the next day by selecting and displaying some of this work to celebrate as you review the previous day's constructions before beginning the new constructions. See examples in Figures 9.10 and 9.11.

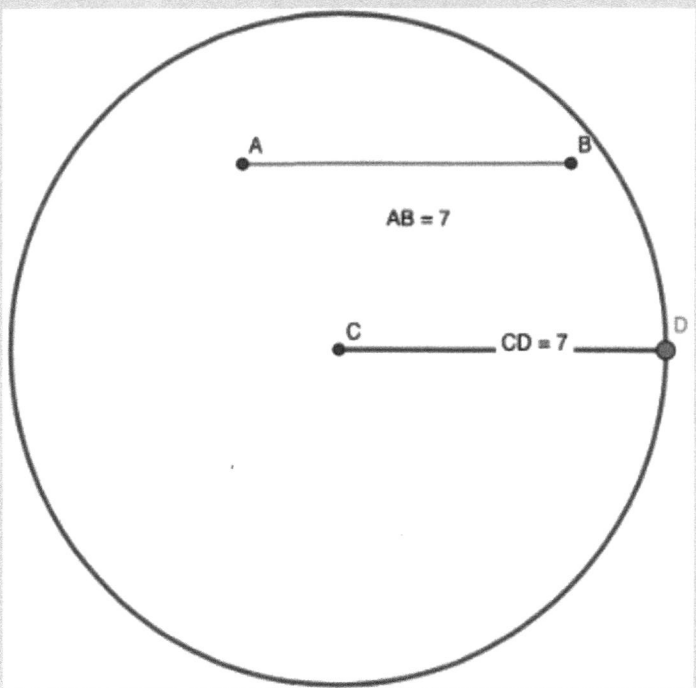

Figure 9.8 Sample duplication of a given line segment construction.

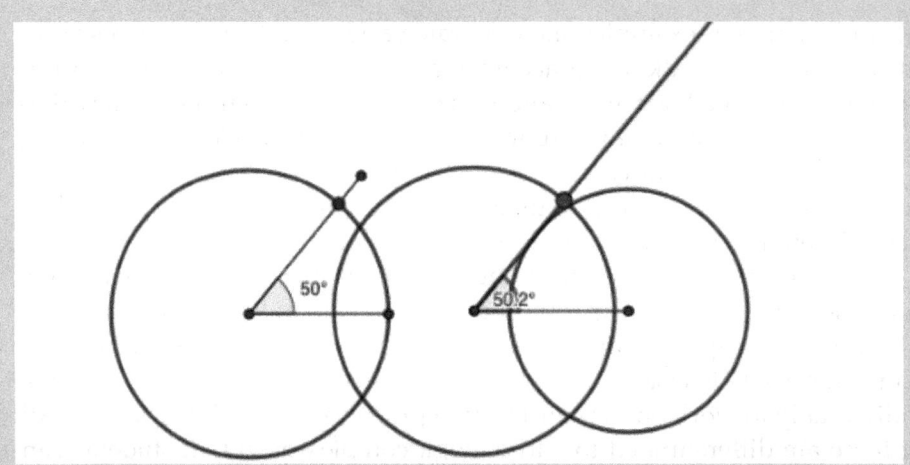

Figure 9.9 Sample copy of a given angle construction.

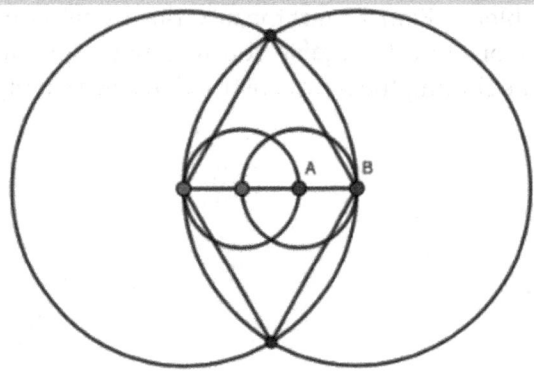

Task 14

Give a brief overview of the steps you took to accomplish the above construction.

Then answer a few of the following questions:
How many triangles did you have to make to accomplish this task? How do you know you are correct?

Answer

I measured the line segment and made two other lines the same size. Then I made the two circles over this new line to find the equilateral triangles. I made two triangles. I know I am right because the two circles ensure all sides are even.

Figure 9.10 Sample student response to challenge 1.

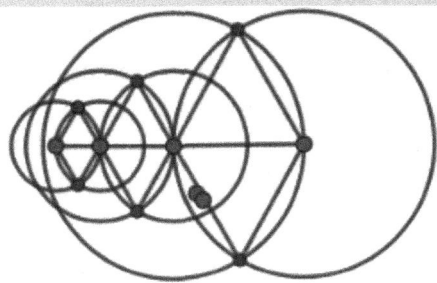

Task 18

Give a brief overview of your challenge, what does someone have to do? What are the steps they need to take to accomplish the above construction.

How did you come up with your construction?

What constructions did you use (#1–4) in your challenge?

Answer

In my challenge I used construction #2 and I had three different segment lengths that connected so that you had to make an equilateral triangle for the first segment then a bigger triangle for the next segment and then an even larger triangle for the last segment.

Figure 9.11 Sample student response to challenge 3.

Whole-Class Task Summary

After students have gone through the exploration and refining stages of the constructions for the day (in this case, constructions 1–4), do a whole-class summary debrief of all the constructions covered that day. Start by asking students what their favorite construction was or maybe what construction they think will be the most useful. If a student is very excited or proud about one of the constructions, ask them if we can show it to the class and celebrate their construction. As the class is more comfortable with each other, you can also ask which construction was the biggest struggle and if anyone would like to share what they first had trouble with and how they overcame it and learned from it. If multiple students are excited about the same constructions, you can use this as an opportunity to show how they can be attempted in slightly different ways to achieve the same result. As long as we are following the same rigorous rules there is room for personality and variants that can show each student's unique thinking about a construction. There is possibly a place to talk about efficiency here, but it is more vital that students feel confident in their own mathematical thinking and ideas.

Advice for Teachers Wanting to Use This Task

Even though the GeoGebra book seems overwhelming, Andrew suggests that you keep it as one book. He considered breaking it up into multiple books or links but he mentioned that his students are always referring back to previous pages. It is helpful for the students to have all the constructions altogether. Andrew also mentioned the importance of working through each construction:

> You need to be able to see the whole scope to see how the later tasks build off the earlier tasks. Also, there are some constructions that you can skip over or use them as extensions. So, it's important to know which constructions are foundational for future constructions.

Andrew discussed that some of the challenge problems can be "quite difficult". He leaves it up to the students to explore those or not. Finally, if you were making constructions on paper to make the constructions you would only make arcs, but in GeoGebra, it will make the entire circle. Because of this, Andrew mentioned how overwhelming the screen can get when the students are working on some of the constructions. To help with this problem, Andrew has students make "their constructions red or green and then leave everything else on the page black". This allows him to quickly look at a sketch to see if students are on the right track.

You can find links to all the technology-enhanced tasks and supplementary videos throughout the book at https://www.tlmtresearch.com/teachingmathtechbook.

10

Using Technology-Enhanced Tasks to Support Student Exploration in Statistics and Probability

Throughout this book, we have shared examples of tasks that are useful when teaching statistics and probability (see Table 10.1). In this chapter, four of the Tech-Math Teachers, Nina, Joel, Shauna, and Lauren, share and unpack their favorite technology-enhanced tasks in this domain. Nina's task is focused on developing a conceptual understanding of the standard deviation. Joel and Shauna's task is focused on using simulations to consider if a situation is likely random or not. Lauren shared a task she designed in which students learn about the importance of random sampling in the context of design-ing a study understanding perceptions of masking related to the COVID-19

Table 10.1 Statistics and Probability Tasks from Prior Chapters

Chapter	Task
2	Figure 2.7 Comparing Data Sets Link 2.2 Valentine Marbles (probability simulations) Link 2.3 Adjustable Spinner Interactive Applet Link 2.6 Monty Hall Game Simulation Link 2.7 Least Squares Regression Lines
5	Link 5.2 Pixar vs. DreamWorks
6	Link 6.3 Making Sense of Box Plots
7	Link 7.4 Exploring Measures of Center

DOI: 10.4324/9781003302285-12

pandemic. The purpose of including these examples is to provide you with not only some great tasks to take and use in your instruction but also insight into the thinking that goes into planning to use the tasks. Each task includes a full implementation plan as well as advice from the Tech-Math Teachers on using it in your classroom. The remainder of this chapter is dedicated to sharing these selected tasks in the domain of probability and statistics.

Example 1: Exploring Standard Deviation

This example, a Desmos activity created by Nina, uses Robert delMas's activity on standard deviation as inspiration. Students use a dot plot that is dynamically linked to a display of their difference from the mean to investigate standard deviation. Nina said that standard deviation is an important concept in her statistics class but is also one that is difficult for her students. Because of that, this is a lesson she always uses as it is well aligned with her learning goals and provides students an opportunity to develop a conceptual understanding of standard deviation.

Description of the Task

This is a Desmos activity titled *Exploring Standard Deviation* (Link 10.1). The task begins with prompting students to recall that the mean is a balancing point and progresses to exploring and making sense of what the standard deviation measures. Specifically, students are asked to consider why the differences between data observations and the mean sum to zero and why this is problematic for calculating the average difference from the mean. Once the procedure for finding standard deviation is introduced, students are asked to consider the relationship between the standard deviation and the difference between individual data points and the mean (i.e., the deviations) by dragging points making sense of how the standard deviation changes as a result. Finally, students are asked how they would explain what the standard deviation measures to a friend.

Link 10.1 Exploring Standard Deviation Desmos Activity

When asked about how this task positions her students as mathematical and statistical **explorers**, Nina explained,

Students begin by dragging data points and observing the changes to the value of the mean and the distance from the mean for each data point. This provides the foundation for all students to explore the connection between the value of the mean and the distance each data point is from the mean. Then later in the activity the students are exploring the effect of moving the data points closer and farther apart on the standard deviation.

The opportunity to drag the data points and see the result is a powerful way for students to explore what the standard deviation is measuring. Reflecting on how the task supports students in **communicating their mathematical and statistical ideas**, Nina noted that the questions throughout required students to communicate their ideas but that the most important part is students' ability to describe what standard deviation is following the task: "After engaging with this task students specifically reference actions they took when dragging points in the task to help describe what standard deviation measures and how that connects to the conceptual definition." In addition, Nina felt that the way this task **builds on students' informal understanding** of standard deviation is powerful because the focus is not on developing formal language. She explained,

> I find that when students are trying to learn something technical, like standard deviation, students tend to focus more on trying to use the correct language to describe the concept which can hinder their understanding. Whereas in this task, when we start with exploring they are not focused on their mathematical and statistical language but instead focused on the movement of the data points, how the value of the standard deviation is changing, and how everything is connected. This gives them an informal way to start digesting what the standard deviation is capturing.

EXPLORING STANDARD DEVIATION: NINA'S IMPLEMENTATION PLAN

Learning Goals

Students will understand the connection between the conceptual and procedural definitions of standard deviation (see definitions provided). Specifically, students will understand the following:

- Why the sum of the distances of the data observations from the mean is 0

- Why the distances must be squared in the standard deviation procedure
- Why there is a square root in the standard deviation procedure
- Why the number of data observations influences the standard deviation
- Why the distance of data observations from the mean influences the standard deviation

Conceptual: Standard deviation measures the average distance from the average.

Procedural: $\sigma = \sqrt{\dfrac{\Sigma(x-\mu)^2}{N}}$

Evidence (i.e., performance goals)

- Students will be able to explain (in writing and/or when sharing their ideas verbally) why the sum of the distances of the data observations from the mean is 0, specifically that the mean is a balancing point and thus the distances of each data observation must balance out.
- Students will be able to explain (in writing and/or when sharing their ideas verbally) why the distances must be squared in the standard deviation procedure by explaining that the standard deviation is an average distance and if the sum of the distances is 0, then we will always get 0 for the standard deviation, which is not true.
- Students will be able to explain (in writing and/or when sharing their ideas verbally) why there is a square root in the standard deviation procedure by explaining that to undo the squaring we must square root.
- Students will be able to apply their understanding of standard deviation to explain why two graphs with the same shape, but a different number of observations have different standard deviations.
- Students will be able to explain (in writing and/or when sharing their ideas verbally) why the distance of data observations from the mean influences the standard deviation.

Instructional Support

This activity is best run with pairs of students, with one computer with internet access to share. The teacher should have a way to project the teacher dashboard to the students so that they can see the (anonymous) responses from the other students.

Prior Knowledge

Students will build on their knowledge of the mean as a balancing point (conceptual understanding of the mean).

Essential Questions

- What does standard deviation measure? How do I calculate standard deviation?
- Why is the sum of the differences between the data observations and the mean always 0?
- What effect does moving one data observation have on the value of the standard deviation?

Task Launch

The first seven pages of the activity are designed to be a warm-up and are used to launch the task. They focus on reviewing the conceptual definition of the mean through exploration. If students are unfamiliar with exploring dynamic data sets in Desmos, begin the activity by displaying screen 1 and explain that each of the blue points is a data observation and that they can be dragged. Drag one point and ask students what they notice. Someone will likely point out that the red x moves, follow that up by noting that the red x is marking the location of the mean. Once students understand what they are looking at, explain that they are going to start the activity by thinking about something we are already familiar with, the mean. Students should be ready to begin engaging with the first screens at that point.

It is recommended to pause after screen 1 and use the anonymize feature of the activity builder to display student responses and discuss the similarities and differences across student responses. On screen 2, it is helpful to remind students that there is a calculator tool at the top they can use. Pause again on screen 7 to compare and contrast how students are explaining why the sum of the deviations of the data observations from the mean is always 0. The goal of the discussion of students' responses on screen 2 is to guide the class to a consensus of the mean as a balance point. Ideally, through the discussion of what the mean is actually measuring, it would have been pointed out that sets of data that have the same mean can look really different. This will drive the need for additional measures to describe the data set. Explain that in this activity we are going to explore ways we might measure the average difference between the data points and the mean. This is a good place to introduce the term *deviation* to describe the difference between a point at the mean, noting that *distance* isn't a great choice as it implies a positive value, but the differences are not all positive. A helpful example is distance from a friend's home: "If I just say I am 5 miles away, that doesn't suggest a direction, whereas −5 could suggest a direction on the street, deviation implies direction". At this point, they will be ready to engage in the remainder of the investigation productively.

Suggestions for Pacing

Since this activity includes an examination of the mean prior to being launched into the exploration of standard deviation, the pacing tool is really important. It is recommended to pace the following screens together: 1, 2–7, 8–10, 11–12, 13–18. Plan to pause for brief whole-class discussions on screens 7, 10, and 12. It is helpful to refer back to screen 12 as an anchor during the whole-class summary discussion of the task.

Pacing 8–10: Pause at screen 10 and recap that we cannot get an accurate average because the differences (i.e., deviations) sum to 0. Since we know that averaging the deviations doesn't work because they sum to zero, ask students what math operations they could use to force the numbers to be positive. Allow students time to think about this with their partner; then generate a list of the different operations that would result in values that still make sense but don't sum to 0 (e.g., absolute value, squaring). Some students will make great suggestions that don't always work (e.g., double the number, multiply by a negative number). Once the ideas are collected, ask which of them would always work. Having considered ways to adjust the differences so that the sum is not zero, students are ready to move on to screens 11 and 12.

Pacing 11–12: Explain that there are multiple ways that statisticians do this. The standard deviation uses squaring and other measures use different operations. For example, MAD (mean absolute deviation) uses the absolute value. Note that we are going to focus on standard deviation in this task. With this in mind, ask students if the sum will continue to be 0 if we square all of the deviations. Let students answer the question, can we average now? Before moving to screen 12, ask students to average the squared deviations (which is 9.25).

On screen 12, discuss how to determine the average of the squared deviations. Then explain that since we decided to square all the deviations, we need to "undo" that. Ask what mathematical operation is the inverse of squaring (i.e., would "undo" the squaring). Explain that a standard deviation of 3.332 means that on average, the points in this data set are 3.332 away from the mean of 3.5. Also show students that the purple-shaded region represents one standard deviation below the mean and one standard deviation above the mean (Figure 10.1).

Pacing 13–18: Students should work through screens 13–18 in pairs. Note: Depending on time, screen 16 can be used as a challenge for groups that are moving faster. Screens 15–17 can be challenging for students. Some questions to pose to advance student thinking when monitoring students' work on these screens include

- What is similar about the purple and green graphs?
- What is different about the purple and green graphs?

Students will typically respond that the purple graph has two additional values. If so, ask where those values are. This is typically enough to advance

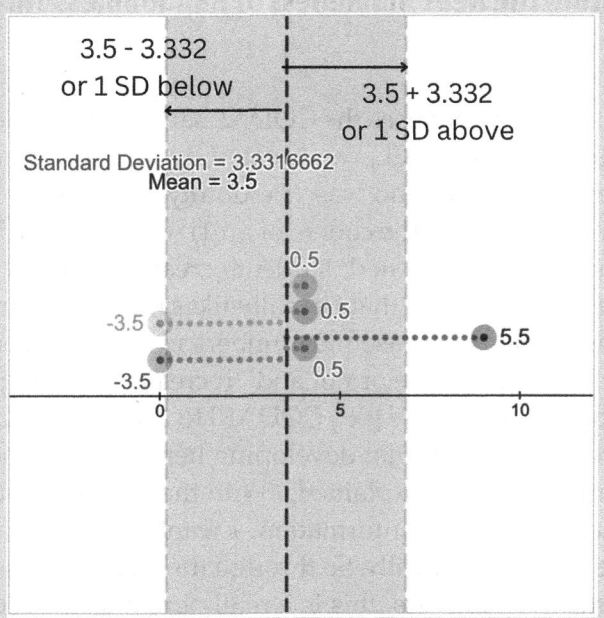

Figure 10.1 Annotating the explanation of standard deviation on screen 12.

the students' thinking. If not, ask them where the mean is and how they know. If more support is needed, you can follow up by asking where the points are in relation to the mean.

Whole-Class Task Summary Discussion

Once the activity is completed, it is important to connect students' conjectures throughout the activity to the formula for standard deviation. One way to do that is to write the formula on the board and ask students to explain each piece and why we do it. Then ask students to explain what the standard deviation measures (i.e., listen for standard deviation is the average distance to the mean). You can also refer back to screen 12 at this time. It is helpful to use examples and questions to illustrate this. For example, you might ask students to consider a test where the teacher says, "The mean is 85, and the standard deviation is 5. What does that tell you about the test scores? What would it indicate if the standard deviation was 2? 10?

Advice for Teachers Wanting to Use This Task

When asked if she had any advice for teachers who would like to use this task, Nina said that first it is crucial to do that task first as a learner. And while working through the task start to think about where you want to pause the students for discussions. She recommended pausing three to four times throughout this task; you might choose different places to pause based on your students' prior experiences with the mean as a measure of the center.

Example 2: Testing the Reasonableness of Randomness Through Simulation

This example, an activity that uses the CODAP sampler to create a simulation to model drug testing in the NFL, was created by Joel and Shauna. Students consider a player, Eric Reid, who was "randomly" selected to participate in drug testing seven times over the course of an 11-week season. To investigate whether this could have happened, students create a sampler and use it to model the situation. Joel shared that he really likes this task as an introduction to probability simulations because the numbers in this situation really make you wonder if this is reasonable or not and "it creates the hunger and appreciation for having this tool to use [i.e., CODAP] to evaluate their intuition". In addition, Shauna emphasized that developing her students' "healthy skepticism" is really important. She explained, "With the rise of social media, that's how they [students] consume information. I want them to challenge everything they come across statistically. So through the lens of developing healthy skeptics, consumers of statistics, this is a really great task". A description of the task including all links follow.

Description of the Task

In this task, the *Eric Reid Probability Modeling Task*, students are introduced to Eric Reid, an NFL player who was hired by the Carolina Panthers after previously being fired (along with his teammate Colin Kaepernick) for publicly kneeling during the National Anthem at NFL games. After considering information regarding NFL drug testing policies, students discuss posts from Eric Reid's social media about being drug tested seven times in 11 weeks. After students discuss their own intuition about the likelihood of this given the Carolina Panthers team data, they will design a simulation in CODAP to model 200 samples of Eric Reid's 11 weeks of random drug test drawings. After compiling the results of each student's simulations, the class will debate the evidence of the simulation and possible reasons for Eric Reid's skepticism.

Link 10.2 The Eric Reid Probability Modeling Task

When considering the ways in which this task positions students as mathematical and statistical **explorers**, Joel noted that the context results in students going through a phase of "this doesn't feel right, but I don't know how to tell if I'm right or not", and then students jump in to use the sampler tool in CODAP to "prove their earlier intuition". This motivation leads them to explore the data themselves as well as create statistics and representations to support their sensemaking. Furthermore, the representations they create are key to supporting students in **communicating their mathematical and statistical ideas**. Joel explained,

Kids want to be able to communicate. They want to build data displays that help communicate what they're already thinking about it. CODAP helps them do that quickly. You can see them get more and more riled up as they see the results of the simulation. They can communicate what I thought was true is actually true, and here's a picture to show that.

When asked about how this task **builds on students' informal understandings**, Shauna shared that students use what they know about mean, shape, variability, spread, box plots, and so on to make sense of the results of the probability simulation. She went on to explain that

the students say, "Oh, like 7 never happened. 6 never happened. We only got two dots at 5." And they start layering in those additional pieces. It really puts some meat behind the meaning of mean and why it's so important. Then they put a box plot on top of that, and like that's your gavel like right there, you know. Now they're using their formal language to debate what they are trying to communicate informally.

THE ERIC REID PROBABILITY MODELING TASK: JOEL AND SHAUNA'S IMPLEMENTATION PLAN

Learning Goals

Note: This task can be used from seventh grade through AP Stat. The specific learning goals might change depending on the prior knowledge you hope students will draw on and the new ideas you plan to incorporate based on the results of the simulation. Some examples are included here.

- Mathematical and Statistical Goals: To design simulations based on real-world limitations to determine the statistical likelihood of an event. Compare probabilities from a model to observed frequencies; if the agreement is not good, explain possible sources of the discrepancy.
- Mathematical Practice Goals: Use appropriate tools (like CODAP) strategically; to construct viable arguments around the likelihood of the event and critique the reasoning of it happening "by chance"
- Critical Goal: Build student confidence in using statistical tools and models to challenge events that seem unjust

Evidence (i.e., performance goals)

- Students will be able to create a simulation in CODAP to model sampling 10 of the 72 Carolina Panthers' players to be drug tested. Within those 72 players, students will clearly designate one to be Eric Reid. Students will use the sampler to simulate 11 weeks of drug testing 10 of the 72 players and determine how many times Eric Reid's name is drawn.
- Students will be able to replicate this simulation 200 (or more) times and compile their results. Using the collected simulations, students will use this model to debate the likelihood of randomly being selected seven times out of 11 weeks drawing on their prior knowledge of
  - central tendency,
  - mean absolute deviation, and
  - box plots and outliers.

Instructional Support

This activity requires student devices (tablets, Chromebooks, or laptops – phones are suboptimal) at a 1-to-1 or 2-to-1 ratio. The teacher will want to project the Google Slides, results of the CODAP simulation, and students' work, which can be shared in the Google Slides.

Prior Knowledge

Students will build on basic probability knowledge to construct a simulation in CODAP that models the real-world scenario behind Eric Reid's "random" drug tests. Upon completion of the simulation, students will use their understanding of one-variable data tools (dot plots, mean/median, mean absolute deviation, etc.) to analyze the results.

Essential Questions

- How can we use established policies and procedures (in this case, NFL drug testing rules and regulations) to set up a simulation to see expected results over many iterations?
- What does it mean when the results of a probability simulation show that real-world results are unlikely?

Task Launch

Note: This is actually a really critical part of the lesson. "Storifying" this activity for students gives them interest and buy-in from early on in the lesson. This part is so important that the Desmos activity linked in the advice section has a video in which Joel does this part for teachers who may not feel as comfortable. What follows are the critical components of the story of Eric Reid, followed by how to transition this story into students thinking about the story from a probabilistic standpoint.

To start the lesson, show the picture of Colin Kaepernick kneeling next to Eric Reid and ask students, "Do you recognize either of the men in this picture?" Chances are that many students will recognize Colin Kaepernick. This allows the opportunity for students to share what they know about him and the teacher to fill in the gaps. The important parts of the story were that he was the starting quarterback for the San Francisco 49ers, led his team to the Super Bowl, and was cut/released from the team after kneeling during the pregame national anthems to protest police brutality against Black men. He was never signed by another NFL team. Recently, he settled a lawsuit against the NFL for unfair labor practices. After hearing the story of Kaepernick, ask students, "Then, we must also know about this other man. Who's he?" Most students cannot name Eric Reid.

At this point, the story transitions to focus on Reid, who was *also* cut after the 2017 season. The difference is that part way through the following season, he was picked up by the Carolina Panthers. After his 11th game with the team, he posted to his Twitter account a picture of his locker after the game, to which the NFL had affixed a notification that he had been randomly selected to complete a postgame drug test, as per league protocol. His only comment on the Twitter post was "Number 7 … 'Random'". Ask students to hypothesize why he typed "Random" in quotes and what those "scare quotes" often indicate about what the writer thinks.

Students usually infer that Mr. Reid doesn't think it was random at all. Follow up with, "Why might he have reason to feel he was being targeted here?" Then ask students what additional information they would like to have to determine whether this seems like a reasonable number of times to be randomly selected. Allow time for students to "turn and talk" with their table partners about what extra information they would like.

When they ask for NFL rules around drug tests, provide it: Each week, the NFL tests 10 players from each team. The Carolina Panthers had 72 players eligible for testing. Then ask students to "turn and talk" again, this time to discuss the following question: "If the Carolina Panthers had 10 players randomly selected out of 72 each week, does it feel reasonable that Eric Reid was selected on 7 out of 11 occasions?"

After students share their feelings with each other, take a poll in the class and allow for some debate and discussion. Typically most students feel that 7 out of 11 seems like "a lot" (some say "too many"), but students lack the tools to demonstrate or prove that. This is where the probability simulation begins, as an attempt to equip students to evaluate their gut feeling that seven times out of 11 games is "too many".

Projecting your laptop, show students the Sampler tool in CODAP. Show how you can add or remove marbles, change how many items are to be collected, and how many samples are to be collected. Then ask, "How can we use this sampler to model this situation and see if getting randomly selected 7 out of 11 games is reasonable?" Provide time for students to discuss and test their ideas with their table partners.

Suggestions for Pacing

It is suggested that after the launch this task is paced in three parts followed by a whole class discussion.

Part 1: After students have had a chance to play around with the sampler, ask more specific questions. For example, have students discuss with table partners: "How can we use this sampler to model the fact that the Panthers had 72 players eligible for testing?" "How can we use it to model that 10 players were selected each week?" "How can we use it to model that Eric Reid was with them for 11 weeks?" Students can enter these values in their own CODAP sampler. Students often get confused about what "select" and "collect" mean when building samplers, but tying it back to the original context is really helpful.

Part 2: Take responses from groups and build a whole-class sampler based on those responses (see Figure 10.2). Notice the first ball in the sampler is labeled "ER" to represent Eric Reid. Then you can choose to either have the students run the simulation many times and compile results (this takes some time) or visit a completed simulation of 200 trials and see the results (Link 10.3).

Link 10.3 CODAP Sampler With 200 Trials Completed

Part 3: Students examine the results (either of their own simulation or of the completed one). They should now feel equipped to use their previous knowledge of one-variable statistics (dot plots, mean/median, box plots/outliers, mean absolute deviation, etc.) to answer the question, "Is seven selections in 11 weeks too many to be truly considered 'random'?" Ask them to use text and images from what they create in CODAP to construct their argument. They can share their arguments in the shared Google Slides. Ask them to share not only what they create in CODAP but also text that explains what they conclude from the results of the simulation. An example of representations students might create are shown in Figures 10.3 and 10.4.

Whole-Class Task Summary

Select and sequence student responses to discuss whether or not we have evidence to support that seven selections in 11 weeks is or is not too many to be considered "random". When selecting and sequencing, it is recommended that you share a wide variety of representations as well as the arguments that go along with them. Most likely the class will come to the

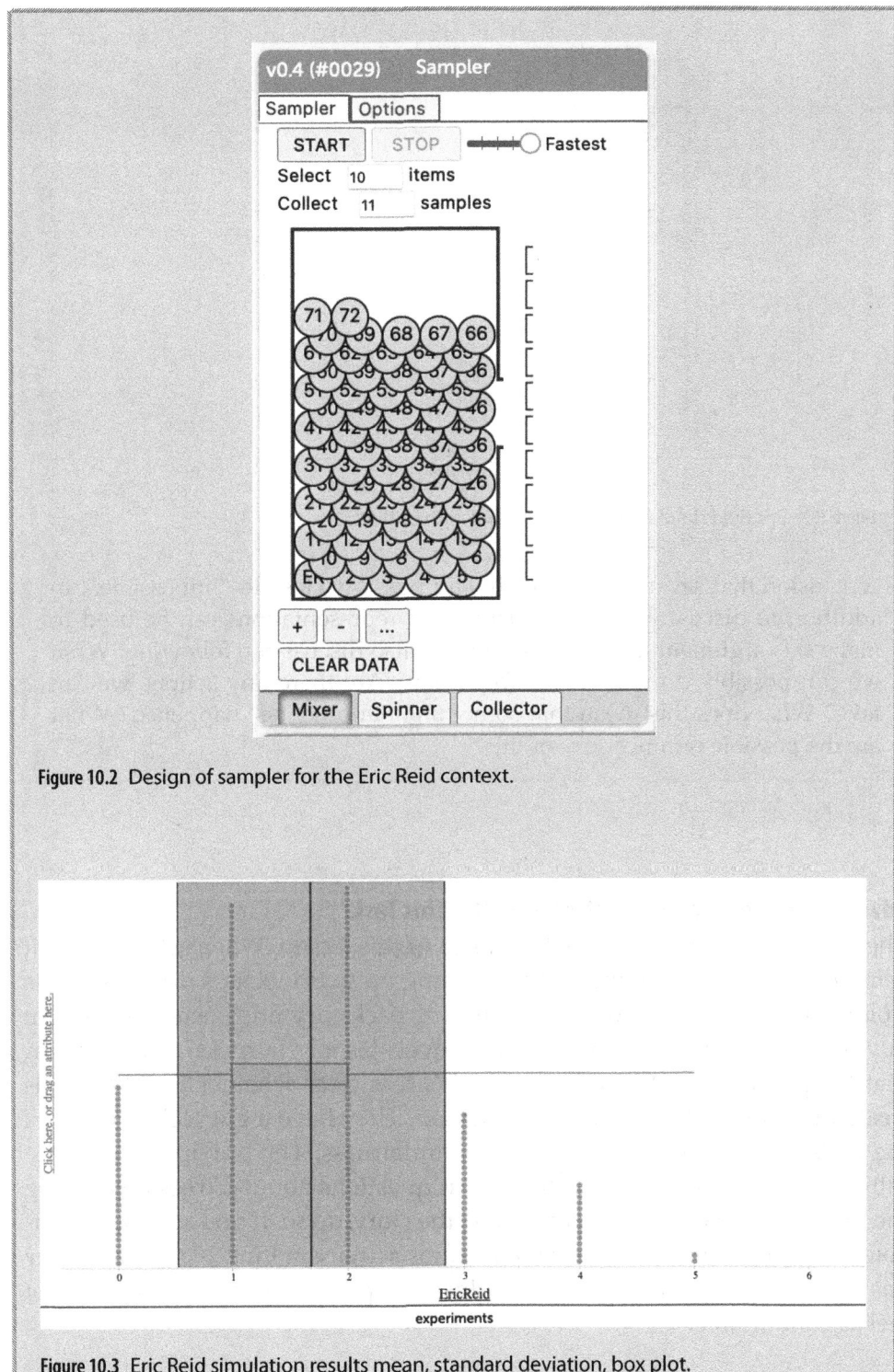

Figure 10.2 Design of sampler for the Eric Reid context.

Figure 10.3 Eric Reid simulation results mean, standard deviation, box plot.

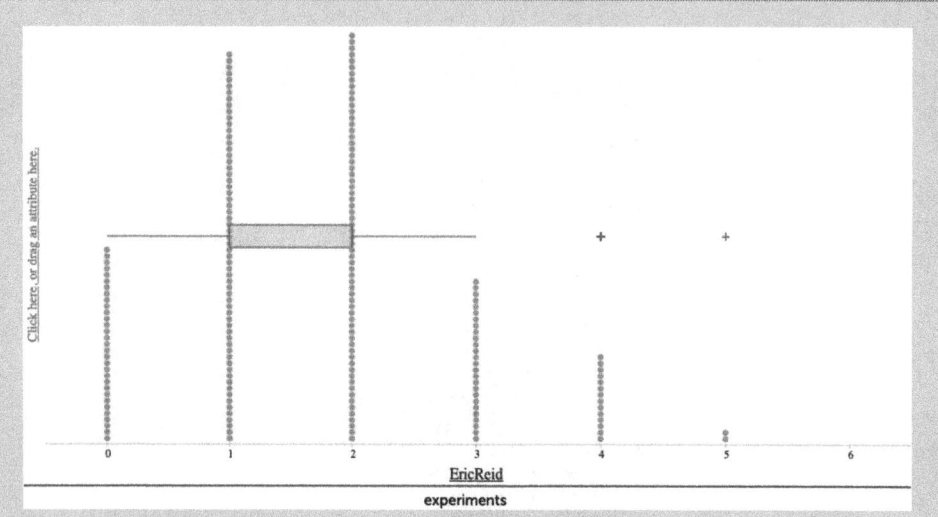

Figure 10.4 Eric Reid simulation results box plot with outliers.

conclusion that seven feels unlikely and possibly close to "impossible". In addition to discussing what statistics and representations can be used to make this argument and how you might also discuss the following: What we can possibly do with that information? Are there any actions we can take? What does this mean that something "impossible" happened? What are the possible ramifications of this?

Advice for Teachers Who Might Want to Use This Task

The story is very important to the launch of this lesson. When setting it up, it is impossible to oversell the "story" leading up to Eric Reid's drug tests. The Colin Kaepernick situation is a necessary backstory and increases student buy-in enormously. Students are huge advocates for "fairness", and it's clear that Kaepernick was treated unfairly. It's less clear whether Eric Reid was treated unfairly at the outset of this lesson. They have a gut feeling but lack the tools to mathematically define that unfairness. The buy-in comes naturally from wanting to answer their own questions about fairness regarding the Eric Reid situation. If you can "sell" the story, do so. If you are uncomfortable doing so, then you can use the Desmos activity in Link 10.4. The activity has a video embedded in which Joel does this part for teachers who may not feel as comfortable.

Link 10.4 The Eric Reid Probability Modeling Task With Video Launch

You might choose to have the students actually run the experiment 200 times themselves and compile results. Joel and Shauna shared that they have found that the coding required to get a new column to calculate the amount of Eric Reid marbles sometimes feels difficult for students. Just like they don't want arithmetic to stand in the way of students being successful at mathematics, neither do they want coding syntax to be an obstacle. That's why they generally provide the completed simulation after setting it up initially with the students' help. ("We just ran it 200 times and counted how often Eric Reid was selected" isn't difficult for students to understand).

Finally, Shauna and Joel shared two different ways you might extend this task. The first is to have students consider what the parameters of the context would need to be for being selected seven times in 11 weeks to not be unusual. Shauna explained,

> So there's just not like an end to this problem. It just continues to grow. What would we have to change? What would we have to do for this not to be statistically significant or to be weird? So I really like that. It allows them to explore in ways that they wouldn't have been able to do if we had not had the CODAP do that.

The second is to capitalize on the students' brainstorming about what actions they might take as a result of their investigation. For example, you could have them write a letter to Roger Goodell, the CEO of the NFL, to support or refute Eric Reid's dispute of being chosen seven times "by chance". Students should justify their position using images and calculations from their simulation and specific evidence of their understanding of central tendency, MAD, box plots, and outliers.

Example 3: Investigating Sample Statistics for Random and Nonrandom Samples

This example, a Desmos activity created by Lauren, uses a large data set, dynamic sampling, and dynamic tape diagrams to consider the relationship between the sample statistics for nonrandom and random samples. Lauren said she liked this task because not only is it well aligned to the learning goals because it walks students through a poorly designed experiment, but it also situates them as the designer – as the person who is manipulating the experiment – so that they can recognize how their own bias influenced the choices they made and ultimately the result of the experiment. A description of the task and the link to the task follow.

Description of the Task

This is a Desmos activity titled *Same Data, Different Results*. Students are asked to go through a warm-up activity in which they look at a scientific article that is being cited as evidence in three newspaper articles and notice and wonder what is different among the claims they are making based on the article. Then, within the activity, students will be asked their opinion about the following question: Do masks reduce the spread of the COVID-19 virus? Students will be sampled for their answers and a statistic will be revealed to the class. The class will then look at whether the statistic is representative of the class as a whole, looking at the summary screen in histogram/box plot form.

Link 10.5 Same Data, Different Results Desmos Activity

When asked about how this task positions her students as mathematical and statistical **explorers** Lauren explained,

> Students are introduced to the idea of randomization through discovery. So they are exploring the need for randomization by creating the headache of understanding that bias exists. Creating the desire for a different way to standardize the way that experiments are made so that you can know that the results are something that you can trust.

The use of dynamic tape diagrams to manipulate sampling procedures, which are then dynamically linked to a headline, serves as a powerful way for students to explore how their biases could affect the outcomes of an experiment. Lauren noted that this was one of the important supports for students **communicating their mathematical and statistical ideas**. She explained,

> I wanted communication occurring across the technology interaction. So as students are working collaboratively and constructing an argument about what changes they were making to the tape diagram, or what changes they were making to the sample in order to make a particular effect occur. As they are changing the dynamic tape diagrams, they're also having to do so in community with a person who may say, "Oh, I want to try this". And for this type of lesson which has these identity constructs in it, that also requires conversation about this critical role.

In addition, Lauren shared that the way that this task **builds on students' informal ideas** about inferential statistics is something she thinks is powerful. She explained,

It allows students to see themselves in the data set. I think that's very important because it informally helps them to draw into the mathematics and to feel like they're a part of what is occurring. It also allows them to make some assumptions about what biased samples do without saying the word bias. It's almost like they get to play puppeteer and in doing so, that informal process of using their inferential statistical knowledge that they're developing helps them to realize the value of randomization and develop healthy skepticism for how we ingest statistics.

SAME DATA, DIFFERENT RESULTS: LAUREN'S IMPLEMENTATION PLAN

Learning Goals

Mathematical and Statistical Goals: To understand how sampling methods affect the results of an experiment, survey, or observation; To understand why randomization is a component of a well-designed experiment.

Interpersonal Goals: To understand that not all statistics about me and others are representative of the entire population

Critical Goals: To build criticality in digesting statistics shared without enough information; To understand how sampling methods can be used to misrepresent a population of people.

Evidence of Student Learning (i.e., performance goals)
- Students will be able to describe variance within a data set.
- Students will be able to use sliders to disproportionately sample a data set and describe how this changes the sample statistic.
- Students will be able to compare and contrast random and nonrandom sampling and sample statistics.
- Students will be able to explain how sampling methods affect the results of an experiment.
- Students will be able to explain why randomization is a component of a well-designed experiment.

Instructional Support

This activity is best run with pairs of students, with one computer with internet access to share. The teacher should have a way to project the teacher dashboard to the students so that they can see the (anonymous) responses from the other students.

Prior Knowledge

Students will build on their knowledge of the statistical process in understanding that all statistics is driven by asking questions. Students will also draw upon their previous conceptual understandings of statistical measures (e.g., mean vs. median) as well as procedural fluency in computing statistical values.

Essential Questions

- How can the sampling strategy a researcher chooses change the conclusions of the study? Is this normal for statistics or is this dangerous?
- Does random sampling always mean that the sample is representative of the population? Why or why not?

Task Launch

This task has an important, prerequisite, warm-up included. In this warm-up, students read a brief excerpt of a scientific article and then compare that to three different news headlines that all reference the scientific article. The context of the scientific article is COVID-19 spread in schools. Students are launched into the task by sharing their own demographic data and being added to the database of answers to the question, "Does wearing a mask reduce the spread of COVID-19?" on screens 4 and 5. Being an individual data point within the larger database places students in a different sense-making position than observing data from outside of the database. On screen 6, students then look at the distribution of characteristics of the participants in the study by gender, race, and geography. This helps them again realize that they are just one individual in a larger database and that there is another representation here. After explaining to students that they, and each of their classmates, are included in that database, explain to students that they will be designing experiments to answer the question, "What percentage of the population agrees that wearing a mask reduces the spread of COVID-19?"

Suggestions for Pacing

The Desmos Classroom pacing tool is really important for this task. It is recommended that you pace the following screens together: 1–3, 4–8, 9–11, and 12. Plan to pause for whole-class discussions on screens 3, 8, and 11. Use screen 12 to facilitate a whole class summary discussion.

On screen 3, look for students who are making sense of the fact that we have three different headlines. This includes recognizing that each headline is pushing a very particular narrative and that each has a storyline that is somewhat attached to the scientific information. Look for students to kind of focus on the nuances of the three while also thinking about how others might generalize about them. This is a great opportunity to get kids to start thinking about assumptions they (or others) might make when they don't have all of the details and/or the role that bias can play in an experiment if we aren't careful. Be careful not to give anything away; it is just an opportunity to get their wheels turning about these sorts of issues before we jump into the task.

On screen 8, after students have had an opportunity to explore with their partner, pause students' work in the activity and use the student view on the dashboard to do a demo of taking samples (Figure 10.5). As you take samples, ask the group, "What is happening when I sample? What do you see changing?" Hopefully you can get to the point where students can answer a question like, "How do statistics depend on our sampling?" Listen specifically for things that are misconceptions about data that students sometimes carry so you can talk about them. One of those is that a statistic is static, it is not something that can change, and that samples are drawn one time, which requires a static statistic. The goal is to get students to recognize and make sense of the fact that when they click the sample button they get a new statistic and that new statistic is dynamically tied to who they are sampling.

Another thing to address on this screen is what the curve represents. Students at this point have never been introduced to a normal curve. So trying to help them make sense of the shape, that this (the middle) is where the bulk of the statistics are going to lie if I sample over and over again, is important. Supporting students in thinking about the curve as the sampling distribution rather than the number of people that agree. An example of selected student responses for such a discussion are shown in Figure 10.6.

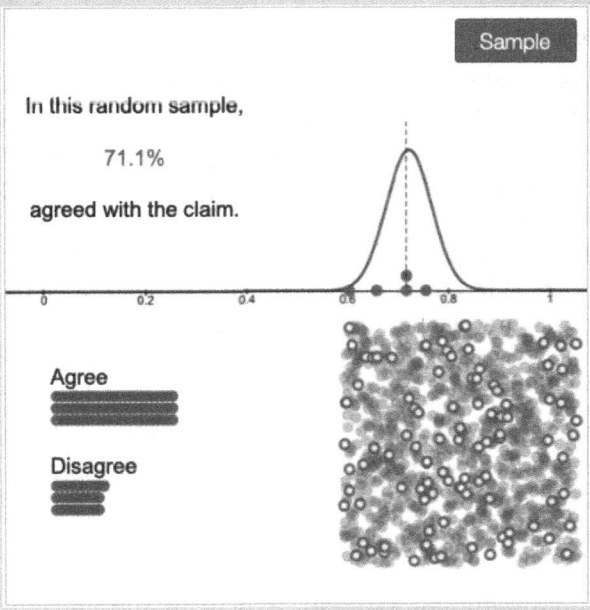

Figure 10.5 Sample screen 8 demonstration.

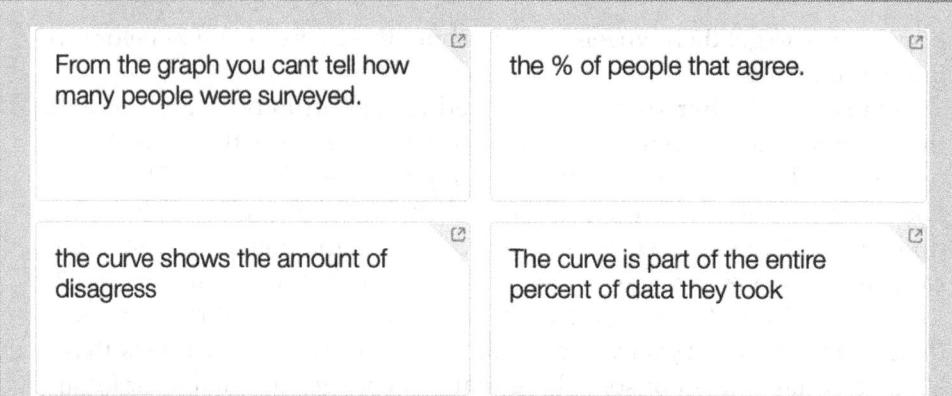

From the graph you cant tell how many people were surveyed.

the % of people that agree.

the curve shows the amount of disagress

The curve is part of the entire percent of data they took

Figure 10.6 Sample student descriptions of what the curve represents.

On screen 11 the root of the conversation should be focused on the fact that we can change the sampling of the populations based on demographics to get different outputs. Ask, "Can you skew the data?" While this is putting them in a position to do something unethical, the goal here is that they recognize that power and the skewing of the data and why that matters. Ultimately, as human beings who design experiments, statistics exist that we should probably question. To get this discussion going, it is recommended that you use screenshots to take a picture of the different tape diagrams that students create to get a really low statistic or a really high statistic and then challenge others to beat it.

Whole-Class Task Summary

It is recommended that you use screen 12 to facilitate a summary of the task as a whole. Once you give students access to this screen using the pacing tool, ask them to just play around with the interaction to make sense of what is happening with a random sample. Then ask them what they notice and wonder. Specifically, use students' noticings to foster conversations about how purposeful sampling can be used to instill bias in data collection and reporting, and that with a random sample, the margin of error on the sample statistic is much smaller.

For example, when looking through her teacher dashboard from the last time she used this task, Lauren excitedly selected the three responses shown in Figure 10.7 and said she would use those for a conversation. She would ask, "Does anybody see something similar or different about these responses?" The goal is to get to a conversation about not only that it is still the same but also why it's staying the same. Lauren went on to explain,

So pulling on the student that said "Where are the statistics coming from?" it is the same sample we had before, but we are no longer

It usually stays around the 70's ?

depending on which groups are represented more, the percentage changes.
why does it stay consistently in the 70s range

I notice that the percent goes up when it is randomly generated compared to the 62%.

I wonder where they are getting these statistics from.

Figure 10.7 Students responses to screen 12.

in control of who we are choosing. So what happens to the statistic when we randomly sample?

In addition, Lauren said she would want to push students to think about why that should matter in the way we do statistics as a community when we are conveying information.

Advice for Teachers Wanting to Use This Task

When asked if she had any advice for teachers who would like to use this task, Lauren focused on the different positions this task puts students in. Not only are they in the position of being statistical doers but statistical doers in unethical ways. With that in mind, Lauren shared the following:

One piece of advice that I think I would give teachers who are trying to use this task is to recognize the value and putting students in the position of being a statistician and a designer of an experiment from a problematic stance. It perturbs so much for them that it supports that need for healthy skepticism in a very different way than asking students to design a well-designed experiment. It's very different. So don't be afraid of allowing students to feel in control. We know that statistics can sometimes be used for harm, and so sometimes you teach students about the harm that something can do by putting them into a position where they actually have to do that thing in a safe space. We don't want to send them out into the world to do real statistics, and then experience how harm can occur in the process. So I think I would really suggest to teachers to hold that space and allow students to really play and engage with the interactions and have real conversations about what that power dynamic in that control is about. Just really be open to it.

An important aspect of a task like this one is for students to reflect on what they have learned. Lauren shared that the next time she uses this task she plans on adding some additional reflection prompts to get students to think about what it felt like to be a part of the sample when an experiment was poorly designed. She would also ask them how it felt to be in the position of being in control of what the statistic would say and why that matters.

 You can find links to all of the technology-enhanced tasks and supplementary video throughout book at https://www.tlmtresearch.com/teachingmathtechbook.

References

Amador, J.M., Glassmeyer, D., & Brakoniecki, A. (2020). Noticing before responding. *Mathematics Teacher: Learning and Teaching PK-12, 113*(4), 310–316. https://doi.org/10.5951/mtlt.2019.0145

Belnap, J.K., & Parrot, A. (2020). Putting technology in its place. *Mathematics Teacher: Learning and Teaching PK-12, 113*(2), 140–146. https://doi.org/10.5951/MTLT.2019.0073

Bonner, E.P. (2021). Practicing culturally responsive mathematics teaching. *Mathematics Teacher: Learning and Teaching PK–12, 114*(1), 6–15. https://doi.org/10.5951/MTLT.2020.0119

Boston, M., Dillon, F., Smith, M.S., & Miller, S. (2017). *Taking action: Implementing effective mathematics teaching practices in grades 9–12.* National Council of Teachers of Mathematics.

Carpenter, T.P., & Lehrer, R. (1999). Teaching and learning mathematics with understanding. In E. Fennema & T.R. Romberg (Eds.), *Mathematics classrooms that promote understanding* (pp. 19–32). Erlbaum.

Cohen, D.K., Raudenbush, S.W., & Ball, D.L. (2003). Resources, instruction, and research. *Educational Evaluation and Policy Analysis, 25*(2), 119–142. https://doi.org/10.3102/01623737025002119

DePeau, E.A., & Kalder, R.S. (2010). Using dynamic technology to present concepts through multiple representations. *Mathematics Teacher, 104*(4), 268–273. https://doi.org/10.5951/MT.104.4.0268

Desmos. (2016). The desmos guide to building great (Digital) math activities. Retrieved from: https://blog.desmos.com/articles/the-desmos-guide-to-building-great-digital-math/

Dick, L., Lovett, J., McCulloch, A., Bailey, N.G., Yalmen Ozen, D., & Cayton, C. (2022). Preservice teacher noticing of students' mathematical thinking in a technology-mediated learning environment. *Submitted to Journal of Mathematics Teacher Education, 29*(3), 129–142. https://doi.org/10.1564/tme_v29.3.02

Dick, L.K., McCulloch, A.W., & Lovett, J.N. (2021). When students use technology tools, what are you noticing? *Mathematics Teacher: Learning and Teaching PK-12, 114*(4), 272–283. https://doi.org/10.5951/mtlt.2020.0285

Dick, T.P., & Hollebrands, K.F. (2011). *Focus in high mathematics: Technology to support reasoning and sense making.* National Council of Teachers of Mathematics.

Doerr, H.M., & Zangor, R. (2000). Creating meaning for and with the graphing calculator. *Educational Studies in Mathematics, 41*(2), 143–163. https://doi.org/10.1023/A:1003905929557

Jackson, K.J., Shahan, E.C., Gibbons, L.K., & Cobb, P.A. (2012). Launching complex tasks. *Mathematics Teaching in the Middle School, 18*(1), 24–29. https://doi.org/10.5951/mathteacmiddscho.18.1.0024

Jacobs, V.R., Lamb, L.L.C., & Philipp, R.A. (2010). Professional noticing of children's mathematical thinking. *Journal for Research in Mathematics Education, 41*(2), 169–202. https://doi.org/10.5951/jresematheduc.41.2.0169

Kenney, R.H. (2014). Investigating a link between precalculus students' use of graphing calculators and their understanding of mathematical symbols. *International Journal for Technology in Mathematics Education, 21*(4), 157–166.

Lappen, G., Fey, J.T., Fitzgerald, W.M., Friel, S.N., & Phillips, E.D. (2009). *Connected mathematics project 2.* Pearson.

McCulloch, A., Lee, H., & Hollebrands, K. (2015). *Preparing to teach mathematics with technology: An integrated approach to algebra.* NC State University.

McCulloch, A.W., Kenney, R.H., & Keene, K.A., (2012). My answers don't math!: Using the graphing calculator to check. *Mathematics Teacher, 105*(6), 464–468. https://doi.org/10.5951/mathteacher.105.6.0464

Moynihan, F., Bejarano, L., & Meyer, D. (2021, February). The Desmos guide to building great (Digital) math activities v2.0. Retrieved from: https://blog.desmos.com/articles/desmos-guide-to-building-great-digital-math-2021/

National Council of Teachers of Mathematics. (2014). *Principles to actions: Ensuring mathematics success for all.* Authors.

National Governors Association Center for Best Practices & Council of Chief State School Officers. (2010). *Common core state standards for mathematics.* Authors.

Papert, S. (1993). *The children's machine: Rethinking school in the age of the computer.* Basic Books.

Pea, R.D. (1985). Beyond amplification: Using the computer to reorganize mental functioning. *Educational psychologist, 20*(4), 167–182. https://doi.org/10.1207/s15326985ep2004_2

Quesada, A.R., & Maxwell, M.E. (1994). The effects of using graphing calculators to enhance college students' performance in precalculus. *Educational Studies in Mathematics, 27*(2), 205–215. https://doi.org/10.1007/BF01278922

Raman, M., & Weber, K. (2006). Key ideas and insights in the context of three high school geometry proofs. *Mathematics Teacher, 99*(9), 644–649. https://doi.org/10.5951/MT.99.9.0644

Ray-Riek, M. (2013). *Powerful problem solving: Activities for sense making with the mathematical practices*. Heinemann.

Resnick, M. (2016, August). Designing for wide walls. Design.blog. Retrieved from https://mres.medium.com/designing-for-wide-walls-323bdb4e7277

Resnick, M., & Silverman, B. (2005, June). Some reflections on designing construction kits for kids. In *Proceedings of the 2005 Conference on Interaction Design and Children* (pp. 117–122). Association for Comuting Machinery, Inc.

Sinclair, M.P. (2003). Some implications of the results of a case study for the design of pre-constructed, dynamic geometry sketches and accompanying materials. *Educational Studies in Mathematics, 52*(3), 289–317. https://doi.org/10.1023/A:1024305603330

Smith, M., Steele, M.D., & Sherin, M.G. (2020). *The 5 practices in practice: Successfully orchestrating mathematical discussions in your high school classroom*. National Council of Teachers of Mathematics.

Smith, M.S., & Stein, M.K. (2011). *Five practices for orchestrating productive mathematics discussions*. National Council of Teachers of Mathematics.

Su, F. (2020). *Mathematics for human flourishing*. Yale University Press.

Suh, J., Roscioli, K., Morrow-Leong, K., & Tate, H. (2022) Transformative technology for equity-centered instruction. *Proceedings of the Society for Information Technology & Teacher Education International Conference* (pp. 1392–1400). Association for the Advancement of Computing in Education (AACE).

Thomas, J., Eisenhardt, S., Fisher, M.H., Schack, E.O., Tassell, J., & Yoder, M. (2015). Professional noticing: Developing responsive mathematics teaching. *Teaching Children Mathematics, 21*(5), 294–303. https://doi.org/10.5951/teacchilmath.21.5.0294

Underwood, J.S., Hoadley, C., Lee, H.S., Hollebrands, K.F., DiGiano, C., & Renninger, K. (2005). IDEA: Identifying design principles in educational applets. *Educational Technology Research and Development, 53*(2), 99–112. https://doi.org/10.1007/BF02504868

Wilson, P.H., Webb, J., Kolb, J., & Wallis, C. (n.d.). The zip line task, Project Lead.